Dynamics in Atmospheric Physics

Lecture Notes for an Introductory Graduate-Level Course

Richard S. Lindzen

Sloan Professor of Meteorology

Dept. of Earth, Atmospheric, and Planetary Sciences
Massachusetts Institute of Technology

CAMBRIDGE
UNIVERSITY PRESS

CAMBRIDGE UNIVERSITY PRESS
Cambridge, New York, Melbourne, Madrid, Cape Town, Singapore, São Paulo

Cambridge University Press
The Edinburgh Building, Cambridge CB2 2RU, UK

Published in the United States of America by Cambridge University Press, New York

www.cambridge.org
Information on this title: www.cambridge.org/9780521361019

© Cambridge University Press 1990

First published 1990
Reprinted 1993
This digitally printed first paperback version 2005

A catalogue record for this publication is available from the British Library

ISBN-13 978-0-521-36101-9 hardback
ISBN-10 0-521-36101-X hardback

ISBN-13 978-0-521-01821-0 paperback
ISBN-10 0-521-01821-8 paperback

Contents

Preface

The following notes have been the basis for an introductory course in atmospheric dynamics which has been taught at Harvard and M.I.T. for the past seven years. The individual chapters were initially intended to correspond to ninety-minute lectures, but, as a result of innumerable changes based on practical demands, the author's predilections, and so forth, this is no longer the case. Some chapters have been reduced while others have been greatly expanded.

Many of the topics covered in these notes may seem somewhat advanced for an introductory course. There are several reasons why they have been included (and why other more traditional topics have been neglected). First, I feel that many topics are considered 'advanced' or 'elementary' for historical reasons and not because they are particularly difficult or easy. The topics I have included do not call on especially advanced mathematical skills; they are, moreover, topics which I believe to be basic to the contemporary study of atmospheric dynamics (*wave-mean flow interaction*, for example). Second, the students who have taken this course at Harvard and M.I.T. have usually had good backgrounds in undergraduate physics and applied mathematics. In many cases, moreover, the students have had some earlier introduction to fluid mechanics. That said, the material in these notes is in many instances conceptually demanding, and students should not feel discouraged if they have difficulty following it. Some topics may require considerable effort.

Because of the background of most of the students, and, in particular, because most of the students have already been introduced to the equations of motion, I have adopted a somewhat unusual approach to the derivation of the equations in Chapter 6. The derivation (which was used by L.N. Howard in his 1960 Woods Hole Lectures on Geophysical Fluid Dynamics – lectures which helped to introduce me to this field) is, I believe, due to Serrin (1959), and, personally, I have always found it thought provoking as well as elegant.

The fact that the equations of motion are not derived until Chapter 6 is discussed in Chapter 1, the Introduction. Here, I wish to note that this ordering is partly due to the nature of the students taking this course at Harvard. Many of these students were planning to work on atmospheric chemistry and radiation rather than on dynamics. Chapters 2 and 3 are designed to show how dynamics plays a rôle in these students' areas of interest. Chapter 4 introduces some (essentially static) force balances (namely, hydrostaticity and geostrophy) which are so intrinsic to the subject that they are actually needed in order to discuss observations, and Chapter 5 deals with observations (much too briefly). Only after describing the observations are the dynamical equations derived. Apart from some technical reasons for this ordering, I have also felt that it serves to emphasize my feeling that atmospheric dynamics is not simply the derivation and application of equations. Rather, one should begin by thinking about nature itself. Consistent with this attitude, I have at two points in these notes (namely, in Chapter 7 on the Hadley circulation and in Chapter 10 on tides) devoted considerable attention to the history of the subject. Hopefully, this will give the student some idea of the context within which certain questions were asked and certain solutions proposed. Stated somewhat differently, some attempt is made to help the student understand why we attempt to solve the problems we do.

It should be stressed that this volume consists in lecture notes for a particular course. As such, the material covered is limited by the time available for that course; the notes are not meant to be a comprehensive text or reference. The use of TEX should not disguise the fact that this book consists simply in the photoreproduction of my class notes. Similarly, I have not included an index (against the advice of the editors) in order to emphasize the fact that this book is not intended as a reference. In mitigation of this omission, the table of contents is sufficiently detailed to permit the reader to find most topics quickly.

Finally, I wish to thank the various students, secretaries, and colleagues who have contributed to the preparation of these notes, most especially, Arlindo DaSilva, who helped develop the exercises, Dr. S. Wofsy, who provided Figures 1–3 of Chapter 3, Prof. R. Dole, who helped assemble Figures 2–15 of Chapter 5, and Prof. R. J. Reed, who provided helpful criticism.

Chapter 1

Introductory remarks

Motion is manifest in the atmosphere in an almost infinite variety of ways including wind chill factors, anomalously cold winters, summer droughts, and so forth. We shall in these lectures somewhat loosely distinguish between motion systems themselves and the things motion systems do. A motion system will be defined by the distributions in space and time of velocity, $\vec{v}$; density, ρ; pressure, p; temperature, T; and constituent densities, ρ_i. Examples of what we will consider to be motion systems are cyclonic storms and stationary planetary scale patterns. The latter are intimately associated with anomalous seasonal weather patterns. Examples of things motion systems do are the reduction of the pole–equator temperature differences to about half of what would be expected from purely radiative considerations, and the creation of maximum column densities of ozone at high latitudes rather than in the tropics as would have been expected on the basis of photochemical equilibrium.

In attempting to discuss this topic in one semester, there is no hope of completeness. Moreover, any atempt to begin at the beginning will barely transcend the beginning itself. Although some background in hydrodynamics is helpful, these lectures are formally self-contained. Also, from the beginning, we will deal with the integration of dynamics with other components of atmospheric physics. Clearly, these lectures cannot be comprehensive. The use of additional references (Houghton, 1977; Holton, 1979; Pedlosky, 1979; Gill, 1982) will be helpful though not essential.

1

We will begin these lectures by considering the rôle of dynamics in several problems where this rôle can be established without detailed reference to the dynamics. The problems we will study are:

1. The rôle of dynamics in simple climate models; and

2. The rôle of dynamics in determining the distribution of chemically active minor constituents.

Having demonstrated the importance of dynamics in the behaviour of the atmosphere, we will turn to the motions themselves. The observations will be reviewed with primary, *but not exclusive*, emphasis on specific motion systems rather than on measurement techniques and problems.

Finally, we will turn to the development and use of the equations of motion. It will be in these lectures that the text material will prove most useful. Our emphasis will *not* be on the multitude of interesting (and less interesting) properties of the equations themselves, but on the use of what the equations tell us in order to understand various phenomena. Among the phenomena we will discuss are:

1. The global distribution of surface easterly and westerly winds;

2. Gravity waves and turbulence in the upper atmosphere;

3. Atmospheric tides;

4. The quasi-biennial oscillation of the stratosphere;

5. Travelling and stationary weather systems;

6. The Gulf Stream.

Don't worry if some of the above terms are unfamiliar to you. They will be explained in the remainder of these notes. Incidentally, the last item in the above list is obviously oceanographic rather than atmospheric (although it involves the action of the atmosphere on the ocean); it is included because it is particularly illuminating.

The above discussions will emphasize physical concepts, with the hope that interested students will delve further into the solution details (insofar as they exist). Finally, formal problems are set at the end of each chapter, and informal questions are proposed throughout the text. Dealing with both is integral to learning the material.

It should be noticed that the next four chapters do not even use the equations of motion. These are not, in fact, introduced until Chapter 6 – following the discussion of observations in Chapter 5. Some hydrostatics relevant to the atmosphere are introduced in Chapter 4, because they are essential to the discussion of observations. Chapters 2 and 3 deal with two problems where dynamics is essential, but where the rôle of dynamics can be delineated without specific reference to details. The hope in introducing the subject in this manner is to avoid the notion that dynamic meteorology is simply the derivation of equations and their subsequent solution.

Chapter 2

Simple energy balance climate models

Supplemental reading:[1]

Budyko (1969)

Held and Suarez (1974)

Lindzen and Farrell (1977)

North (1975)

Sellers (1969)

We will consider energy balance climate models because they are the simplest models wherein the interactions of radiation (including the effect of snow on albedo) and dynamic heat transport can be considered. In the above references some attempt is made to justify the realism of the models. Although this is probably worth thinking about, we are here only concerned with the illustrative aspects rather than detailed realism.

---

[1]A complete list of references is given at the end of this book. Those references that are particularly useful to a given chapter are listed at the beginning of that chapter. Sometimes specific pages and/or chapters will be noted.

These models are typically characterized as follows:

1. Only latitude dependences are considered; that is, the models are spatially one-dimensional (though time dependence is also sometimes considered).

2. Global energy budgets are assumed to be expressible in terms of surface temperatures.

3. Planetary albedo is taken to depend primarily on ice and/or snow cover or the lack thereof.

4. The convergence of dynamic heat fluxes is generally represented by either a simple diffusion law or by a linear heating law wherein local heating is proportional to deviations of the global mean temperature from the local surface temperature. *The primary feature of the heat transport is that it carries heat from warmer to colder regions.* Both of the above representations do this.

5. Generally, only annual mean conditions are considered.

The starting point for such models is an equation of the form

$$C\frac{\partial T(x,t)}{\partial t} = \text{incoming solar radiation}$$
$$-\text{infrared cooling}$$
$$-\text{divergence of heat flux}, \quad (2.1)$$

where C is some heat capacity of the atmosphere-ocean system, T the surface temperature (°C), t is time, and $x = \sin\theta$, where θ is the latitude. It is somewhat more convenient to deal with x rather than θ.

Under the assumption that the *total* global energy budget can be expressed in terms of the surface temperature, the first term on the right-hand side of Equation 2.1 is generally taken to be the total insolation as might be determined by a satellite above the atmosphere. It is typically written as

$$\text{incoming solar radiation} = Qs(x)\mathcal{A}(T), \quad (2.2)$$

where Q is one quarter of the solar constant (Why?) and $s(x)$ is a function whose integral from the equator to the pole is unity and which represents the annually averaged latitude distribution of incoming radiation. This function is discussed in Held and Suarez (1974). Finally, $\mathcal{A}(T)$ is 1 minus the planetary albedo; $\mathcal{A}$ is allowed to depend on temperature. In most simple climate models, a temperature T_s is identified with the onset of ice (snow) cover such that for $T > T_s$ there is no ice (snow), and for $T < T_s$ there is. The most important change in $\mathcal{A}$ is due to T passing through T_s. We will specify $\mathcal{A}$ more explicitly later.

Again under the assumption that global energy budgets can be expressed in terms of surface temperature, one writes

$$\text{infrared cooling} = I(T). \tag{2.3}$$

The justification for Equation 2.3 is that temperature profiles have more or less the same shape at all latitudes. Hence cooling, which depends on the temperature at all levels, ought to be expressible in terms of surface temperature, since the temperature at all levels is related to the surface temperature. In fact, temperature profiles at different latitudes are somewhat different (*viz.* Figure 2.1). Moreover, Held and Suarez (1974) have shown that 500mb temperatures correlate better with infrared emission than do surface temperatures. Nevertheless, it is the surface temperature which relates to the formation of ice, and which therefore must be used in simple climate models. The fact that total infrared emission is not perfectly related to surface temperature is merely an indication that a significant portion of the emitted radiation originates in the atmosphere. Similarly, not all of the incoming radiation is absorbed at the surface; in practice, some of the incoming radiation is *not* directly involved in the surface energy budget.

As a rule, models based on Equation 2.1 take little account of clouds and cloud feedbacks. In truth we hardly know how to include such feedbacks. It is probably impossible in such a simple model. However, to the extent that clouds can be specified in terms of latitude and surface temperature, their effects on incoming radiation and on infrared emission can be included in Equations 2.2 and 2.3.

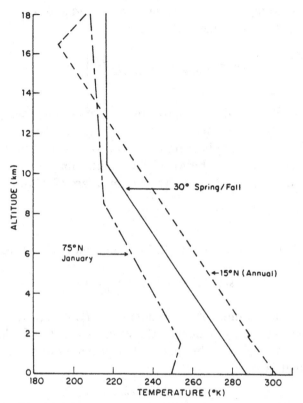

Figure 2.1: Vertical temperature profiles for various latitudes. (From U.S. Standard Atmosphere Supplements, 1966.)

The divergence of atmospheric and oceanic heat flux must also be expressed in terms of an operator on surface temperature; that is,

$$\text{div flux} - F[T], \tag{2.4}$$

where F is some operator. Usually F is a linear operator, although Held and Suarez (1974) and North (1975) have also considered nonlinear operators as suggested by Green (1970) and Stone (1973). The common choices for $F[T]$ are a linear relation first suggested by Budyko (1969):

$$F[T] = C \ (\bar{T} - T), \tag{2.5}$$

where $\bar{T}$ is the average of T over all latitudes[2], and a diffusion law (first used in this context by Sellers (1969)):

$$F[T] = \frac{\partial}{\partial x}\left[(1-x^2)D\frac{\partial T}{\partial x}\right],\qquad(2.6)$$

where C and D in (2.5) and (2.6) are constants; they are generally chosen to simulate some features of the existing climate.

The most common application of (2.1) involves assuming a steady state and seeking a relation between the equilibrium position of the ice line and the solar constant. Usually, one linearizes (2.3) to obtain

$$I = A + BT,\qquad(2.7)$$

and replaces T by I as defined in (2.7). Equation 2.1 becomes[3]

$$Qs(x)\mathcal{A}(I) - I + F^*[I] = 0.\qquad(2.8)$$

One identifies the ice line with a temperature T_s, or equivalently $I_s = A + BT_s$. We shall use x_s to identify the value of x at $I = I_s$. Moreover, variations in $\mathcal{A}$ are taken to be due solely to whether or not there is an ice surface, that is,

$$\mathcal{A} = \mathcal{A}(x, x_s).\qquad(2.9)$$

'C' or 'D' in (2.4) will be chosen so that for the present climate $T = T_s$ at the present annually averaged value of x_s (i.e., $x_s \simeq 0.95$). Obtaining the dependence of x_s on Q (or more conveniently, the dependence of Q on x_s) is straightforward. If we write

$$I = Q\tilde{I}(x),\qquad(2.10)$$

and assume F to be a linear operator, then we may divide (2.8) by Q, yielding

$$-F^*[\tilde{I}] + \tilde{I} = s(x)\mathcal{A}(x, x_s).\qquad(2.11)$$

---

[2]A general constraint on $F[T]$ is that its integral over the globe be zero (Why?).
[3]$F^* \equiv \frac{1}{B}F$

For any choice of x_s we may solve (2.11) for $\tilde{I} = \tilde{I}(x, x_s)$. It is now a trivial matter to obtain the solar constant (or, equivalently, Q) as a function of x_s. We already have

$$I(x_s) = I_s. \tag{2.12}$$

But since

$$I(x_s) = \tilde{I}(x_s, x_s)Q, \tag{2.13}$$

by combining (2.12) and (2.13) we have

$$\frac{Q}{I_s} = \frac{1}{\tilde{I}(x_s, x_s)}, \tag{2.14}$$

which is the desired relation.

Normally, we expect advancing ice (decreasing x_s) to be associated with decreasing Q. Such a situation is generally stable in the sense that the time-dependent version (2.1) indicates that perturbations away from the equilibria defined by (2.14) decay in time. This stability is easy to understand intuitively. If, for example, one decreased x_s while holding Q constant, then Q would be larger than needed for that value of x_s and the resulting warming would cause x_s to increase. This is, in fact, the situation when we do *not* have transport. Surprisingly, the introduction of transport always leads to some values of x_s where decreasing x_s is associated with increasing Q – an unstable situation leading to an ice covered earth (at least in the context of the simple model).

We shall examine how this occurs under particularly simple conditions. First we shall use (2.5) for $F[T]$. Next we shall take the following expression for $\mathcal{A}$:

$$\mathcal{A} = \alpha \text{ for } T < T_s$$

$$\mathcal{A} = \beta \text{ for } T > T_s. \tag{2.15}$$

Common choices for α and β are:

$$\alpha = 0.4$$

$$\beta = 0.7. \tag{2.16}$$

For T_s we, for the moment, take $T_s = -10°C$, and for the constants A and B in (2.7) we use $A = 211.1\,\mathrm{Wm^{-2}}$, and $B = 1.55\,\mathrm{Wm^{-2}(°C)^{-1}}$. Hence, $I_s = 195.6\,\mathrm{Wm^{-2}}$. For $s(x)$ we use the annual average function as approximated by North (1975):

$$s(x) \approx 1 - 0.241(3x^2 - 1). \tag{2.17}$$

For the present solar constant,

$$Q = 334.4\,\mathrm{Wm^{-2}}.$$

For the above choices, straightforward analytic solutions exist. Let us begin by neglecting all transport. Equation 2.8 becomes

$$Qs(x)\mathcal{A}(x, x_s) - A - BT = 0,$$

or

$$T = \frac{Qs(x)\mathcal{A}(x, x_s) - A}{B}. \tag{2.18}$$

This radiative equilibrium value of T depends only on the local radiative budget. Its distribution is shown in Figure 2.2 as is the observed distribution. Note that the observed gradients are much smaller than in the equilibrium distribution. The value of Q associated with x_s is not unique. Any value of Q less than the value needed for $T(x_s) = T_s$ with $\mathcal{A} = \alpha$, and greater than the value needed for $T(x_s) = T_s$ with $\mathcal{A} = \beta$, is consistent with x_s. This leads to the two curves for $Q(x_s)$ shown in Figure 2.3:

$$Q_+(x_s) = \frac{A + BT_s}{s(x_s)\alpha}, \tag{2.19}$$

$$Q_-(x_s) = \frac{A + BT_s}{s(x_s)\beta}. \tag{2.20}$$

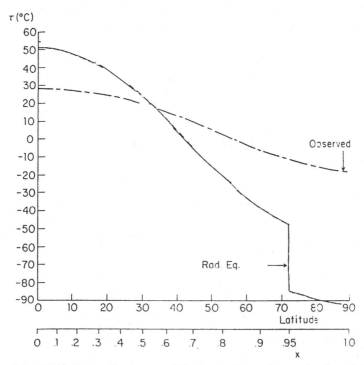

Figure 2.2: $T(\phi)$ for radiative equilibrium. Also shown is the observed $T(\phi)$. For reference purposes, x as well as ϕ is shown. Taken from Lindzen and Farrell (1977).

It will be left to the student to figure out how x_s will vary as Q is changed, but it is evident that for both (2.19) and (2.20) that decreasing Q leads to decreasing x_s and *vice versa*. (Note that $s(x_s)$ decreases monotonically with x_s.)

Introducing transport via Equation 2.5 does not eliminate the ambiguity in Q. A device for eliminating the ambiguity is to choose

$$A(x_s) = \frac{\alpha + \beta}{2}. \tag{2.21}$$

This device turns out to be almost equivalent to introducing a very small conductivity which, in turn, renders T continuous at x_s (*viz.* Figure 2.3).

Now

$$F^*[\tilde{I}] = (C/B)(\bar{\tilde{I}} - \tilde{I}) \qquad (2.22)$$

and

$$\bar{\tilde{I}} = \int_0^1 s(x)\,\mathcal{A}(x, x_s)\,dx \qquad (2.23)$$

(Why?).

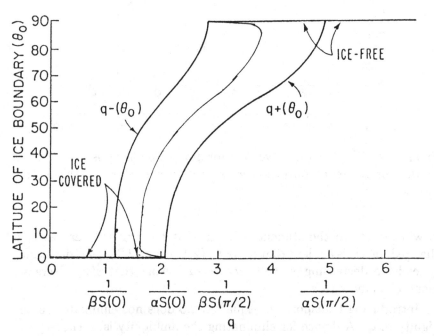

Figure 2.3: Variation of q ($q \equiv Q/I_s$) vs. θ_s ($\sin^{-1} x_s$) for radiative equilibrium. The curve q_- represents $T = T_s$ for the ice-free side of x_s, while q_+ represents $T = T_s$ for the ice-covered side of x_s. Also shown is the curve obtained with a very small amount of diffusive heat transport. From Held and Suarez (1974).

Note that θ_0 in this figure corresponds to θ_s in the text.

Evaluating (2.23) (using (2.15) and (2.17)) we get

$$\bar{\bar{I}} = (\beta - \alpha)(1.241\,x_s - .241\,x_s^3) + \alpha. \tag{2.24}$$

Substituting (2.22) in (2.11), we get

$$\tilde{I} = \frac{\frac{C}{B}\bar{\bar{I}}(x_s) + s(x)\,\mathcal{A}(x, x_s)}{1 + \frac{C}{B}}. \tag{2.25}$$

Equation 2.25 allows us to determine C such that $\tilde{I}(.95) = \tilde{I}_s$ for present conditions ($C/B = 2.45$ is obtained). Assuming this value of C remains constant as Q varies, we then get from (2.14)

$$Q = \frac{(1 + \frac{C}{B})(A + BT_s)}{\frac{C}{B}\bar{\bar{I}}(x_s) + s(x_s)(\frac{\alpha + \beta}{2})}. \tag{2.26}$$

Examining the denominator of (2.26) in detail we get

$$\begin{aligned}
\text{den} \;=\; & \frac{\beta + \alpha}{2} \times 1.241 + \alpha\frac{C}{B} \\
& + \frac{C}{B}(\beta - \alpha) \times 1.241\,x_s \\
& - \frac{(\beta + \alpha)}{2} \times .723\,x_s^2 - \frac{C}{B}(\beta - \alpha) \times .241\,x_s^3. \tag{2.27}
\end{aligned}$$

When $C = 0$, the denominator decreases as x_s increases, as already noted. But, when $C \neq 0$, there always exists some neighbourhood of $x_s = 0$ where the linear term in (2.27) dominates, and the denominator *increases* as x_s increases. This leads to the distribution of Q versus x_s shown in Figure 2.4. Two features should be noted in Figure 2.4, both being due to the existence of horizontal heat transport:

1. A much smaller value of Q is needed for ice/snow to onset at all. This represents the stabilizing effect of transport.

2. There now exists some minimum Q, below which the climate will unstably proceed to an ice/snow covered earth. This represents the destabilizing effect of transport.

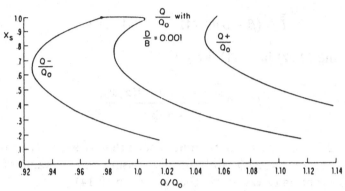

Figure 2.4: Equilibrium ice line position x_s as a function of normalized solar constant Q/Q_0 (where Q_0 is the current value of the solar constant) when Budyko-type heat transport is used. The curve Q_-/Q_0 corresponds to $T = T_s$ on the equatorward side of the ice line, while Q_+/Q_0 corresponds to $T = T_s$ on the poleward side. The single remaining curve results from adding a small amount of diffusion to the Budyko-type transport.

Both the above effects result very generally from the sharing of heat between low and high latitudes. Clearly, heat transport from low latitudes initially inhibits the onset of ice/snow at the poles. However, as the ice/snow line advances, the transport of heat out of warmer regions cools these regions to such an extent that Q must actually be increased to keep up with further advances. This situation is clearly unstable. We shall refer to the percentage Q must be reduced from its present value to reach instability as the 'global stability'. The results in Figure 2.4 correspond to a global stability of only $\sim 2\%$. To be sure, the solar constant might not vary this much. However, Q can be viewed as a general measure of global heating. Changes in Q can be simulated by changes in I and/or $\mathcal{A}$, for example.

As will be seen in the exercise, the above estimate of *global stability* is hardly firm, but our only interest at this point is in the general rôle of heat transport.

Exercise

2.1 Let $\alpha = 0.45$. Keep A and B as given in the text.

1. Determine β such that $\bar{T}$ remains unchanged.

2. Determine C for the above choices of α and β.

3. Compute $Q(x_s)$.

4. Discuss any differences between these results and those obtained for $\alpha = .4$, $\beta = .7$. In particular, how has the global stability changed and why?

Chapter 3

Effect of transport on composition

Supplemental reading:

Holton (1979), pp. 40–4

Houghton (1977), pp. 81–2, 55–8

3.1 General considerations

Consider a chemical constituent, i, with density $\rho_i(z, \theta)$ and a photo-chemical equilibrium distribution $\overline{\rho_i}(z, \theta)$. Let us consider an idealized situation where, in the absence of transport,

$$\frac{\partial \rho_i}{\partial t} = \alpha(z, \theta)(\overline{\rho_i}(z, \theta) - \rho_i(z, \theta)), \tag{3.1}$$

where z is altitude and θ is latitude. Equation 3.1 may be considered as a highly idealized description of ozone. Figure 3.1 shows the distribution of $\overline{\rho_i}(z, \theta)$ (for ozone); Figure 3.2 shows the observed distribution; Figure 3.3 shows the photochemical relaxation time (α^{-1}). A comparison of Figures 3.1 and 3.2 shows that atmospheric ozone is, in most regions, not in photochemical equilibrium. A notable excep-

16

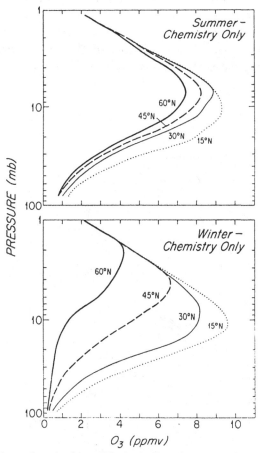

Figure 3.1: Photochemical equilibrium distributions of ozone mixing ratio with pressure at various latitudes for winter and summer. Note that this and the following three figures were prepared by S. Wofsy in 1980. They are not state-of-the-art calculations; this doesn't particularly matter for the crude arguments of this chapter.

tion is the tropical upper stratosphere. The differences between the observed and photochemical equilibrium distributions become particularly clear when we focus on column densities (i.e., the total ozone per unit area above a given point). These are shown in Figure 3.4. We see that the equilibrium distribution has a maximum over the equator, and decreases toward the poles in both summer and winter – with the winter minimum being much deeper. The observed distribution has a minimum at the equator and rises toward the poles, with the winter maximum being greater than the summer maximum.

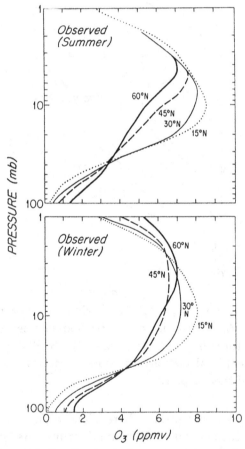

Figure 3.2: Observed distributions of ozone mixing ratio with pressure at various latitudes for winter and summer.

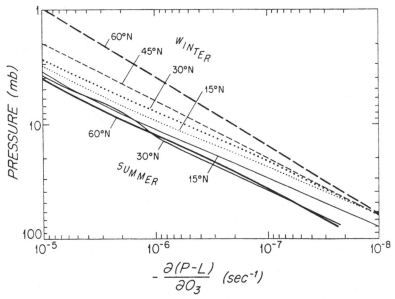

Figure 3.3: Photochemical relaxation rate for ozone as a function of pressure for various latitudes. This rate is estimated by differentiating the difference between ozone loss and production with respect to changes in ozone.

3.1.1 Equations of continuity

We wish, now, to examine the rôle of a large-scale motion field in causing ρ_i to differ from $\overline{\rho_i}$. To do this we must introduce the equation of mass continuity. Consider a fixed element of volume in cartesian coordinates (*viz.* Figure 3.5):

$$\frac{\partial \rho}{\partial t} \delta x\, \delta y\, \delta z \quad =$$

$$\left[\rho u - \frac{\partial}{\partial x}(\rho u)\frac{\delta x}{2}\right]\delta y\, \delta z \;-\; \left[\rho u + \frac{\partial}{\partial x}(\rho u)\frac{\delta x}{2}\right]\delta y\, \delta z$$

$$+\left[\rho v - \frac{\partial}{\partial y}(\rho v)\frac{\delta y}{2}\right]\delta x\, \delta z \;-\; \left[\rho v + \frac{\partial}{\partial y}(\rho v)\frac{\delta y}{2}\right]\delta x\, \delta z$$

$$+\left[\rho w - \frac{\partial}{\partial z}(\rho w)\frac{\delta z}{2}\right]\delta x\, \delta y \;-\; \left[\rho w + \frac{\partial}{\partial z}(\rho w)\frac{\delta z}{2}\right]\delta x\, \delta y.$$

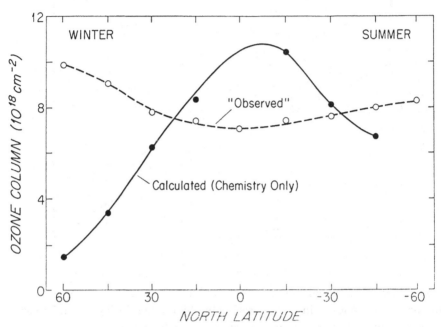

Figure 3.4: Observed and calculated (on the basis of photochemical equilibrium) distributions of ozone column density with latitude in the northern hemisphere for both winter and summer.

As $\delta x, \delta y, \delta z \to 0$, we get

$$\frac{\partial \rho}{\partial t} = -\nabla \cdot (\rho \vec{u}).$$ (3.2)

In a similar manner (3.1) may be generalized to

$$\frac{\partial \rho_i}{\partial t} + \nabla \cdot (\rho_i \vec{u}) = \alpha(\overline{\rho_i} - \rho_i).$$ (3.3)

3.2   4-box transport model

We next wish to apply (3.2) and (3.3) to a very simplified geometry where $\vec{u}$ is specified (*viz.* Figure 3.6). Each of the four boxes has the same basic mass (i.e., $p_2 = p_1 - \Delta p$; $p_3 = p_2 - \Delta p$). At each interface we will assume the velocity to have a characteristic magnitude V (at vertical surfaces) or W (at horizontal surfaces). Finally, we assume that each box can be characterized by single values of α_j, $(\rho_i)_j$, $(\overline{\rho_i})_j$, and ρ_j, where j = box number (Note, ρ_j refers to the mean density of air in the jth box, $(\rho_i)_j$ to the density of the ith constituent in the jth box, and $(\overline{\rho_i})_j$ to the photochemical equilibrium density of the ith component in the jth box.) This approach is, of course, extremely crude, but it is adequate for illustrative purposes.

Let us first integrate Equation 3.2 over box 1 (we are assuming a steady state where $\frac{\partial}{\partial t} = 0$):

$$\int_{\text{box 1}} \nabla \cdot \rho \vec{u} \, dy \, dz = \int_{\text{perimeter of box 1}} \rho u_n \, dl$$
$$= \rho_1 V \Delta H_1 - \rho_4 W L = 0.$$

This implies

$$\rho_1 V \Delta H_1 = \rho_4 W L = M. \tag{3.4}$$

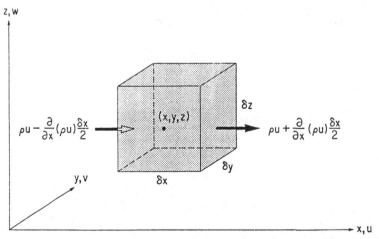

Figure 3.5: Schematic depiction of mass flow and continuity.

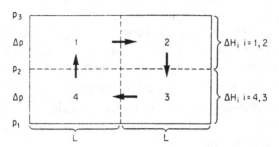

Figure 3.6: 4-box geometry for studying effect of transport on a chemically active constituent.

Making use of hydrostaticity,[1]

$$\Delta H_1 \cong \frac{\Delta P}{\rho_1 g}$$

so (3.4) becomes

$$\frac{\Delta P}{g} V = \rho_4 W L = M. \tag{3.5}$$

More generally, the mass flux across each interface must equal M. (As a practical matter, W may have to be considered different according to whether it is going up or down.)

Integrating Equation 3.3 over box 1 we get

$$(\rho_i)_1 V \Delta H_1 - (\rho_i)_4 W L = \alpha_1 ((\overline{\rho_i})_1 - (\rho_i)_1) L \Delta H_1 \tag{3.6}$$

and, using (3.5),

$$\frac{(\rho_i)_1}{\rho_1} M - \frac{(\rho_i)_4}{\rho_4} M = \alpha_1 \frac{\Delta P}{g} L \left(\frac{(\overline{\rho_i})_1}{\rho_1} - \frac{(\rho_i)_1}{\rho_1} \right)$$

or, more generally,

$$\left(\frac{\rho_i}{\rho} \right)_j M - \left(\frac{\rho_i}{\rho} \right)_{j-1} M = \alpha_j \frac{\Delta P}{g} L \left(\left(\frac{\overline{\rho_i}}{\rho} \right)_j - \left(\frac{\rho_i}{\rho} \right)_j \right)$$

---

[1]If the reader doesn't know what this is, it is discussed in Chapter 4.

or, equivalently,

$$R\left[\left(\frac{\rho_i}{\rho}\right)_j - \left(\frac{\rho_i}{\rho}\right)_{j-1}\right] = \alpha_j\left(\left(\frac{\overline{\rho_i}}{\rho}\right)_j - \left(\frac{\rho_i}{\rho}\right)_j\right) \qquad (3.7)$$

where

$$R = \frac{Mg}{\Delta P L} = \frac{V}{L}$$

and

$$j = 1, 2, 3, 4$$
$$j - 1 = 4, 1, 2, 3$$

(i.e., j is a cyclic index where $j = j + 4$).

Several important points are to be noted concerning (3.7):

1. The dimension of both R and α_j is $1/[T]$ (i.e., $1/\text{time}$).

2. The left-hand side of (3.7) represents the rate at which advection is acting to eliminate differences in $(\frac{\rho_i}{\rho})$ between adjacent boxes. Note that advection acts to homogenize the mixing ratio of constituent i, $(\frac{\rho_i}{\rho})$, rather than its density, ρ_i.[2]

---

[2]A somewhat more elegant approach to this feature can be obtained directly from Equations 3.2 and 3.3 Rewrite $\rho_i = \frac{\rho_i}{\rho}\rho$. Then

$$\frac{\partial \rho_i}{\partial t} + \nabla \cdot (\rho_i \vec{u}) = \rho\frac{\partial}{\partial t}\left(\frac{\rho_i}{\rho}\right) + \frac{\rho_i}{\rho}\frac{\partial \rho}{\partial t}$$

$$+ \rho\vec{u} \cdot \nabla\left(\frac{\rho_i}{\rho}\right) + \frac{\rho_i}{\rho}\nabla \cdot (\rho\vec{u})$$

$$= \rho\left\{\frac{\partial}{\partial t}\left(\frac{\rho_i}{\rho}\right) + \vec{u} \cdot \nabla\left(\frac{\rho_i}{\rho}\right)\right\} = \alpha(\overline{\rho_i} - \rho_i)$$

and

$$\frac{\partial}{\partial t}\left(\frac{\rho_i}{\rho}\right) + \vec{u} \cdot \nabla\left(\frac{\rho_i}{\rho}\right) = \alpha\left(\frac{\overline{\rho_i}}{\rho} - \frac{\rho_i}{\rho}\right).$$

3. The right-hand side of (3.7) represents the rate at which chemistry is acting to bring $\left(\frac{\rho_i}{\rho}\right)$ to its equilibrium value, $\left(\frac{\overline{\rho_i}}{\rho}\right)$.

4. The non-dimensional parameter R/α represents the balance in the competition between the two processes described in items (2) and (3). When $R/\alpha \gg 1$, there is a tendency for $\left(\frac{\rho_i}{\rho}\right)_j$ to approach $\left(\frac{\rho_i}{\rho}\right)_{j-1}$, whereas when $R/\alpha \ll 1$, there is a tendency for $\left(\frac{\rho_i}{\rho}\right)_j$ to approach its chemical equilibrium value.

Our object is to solve (3.7) for $\left(\frac{\rho_i}{\rho}\right)_j$. This is facilitated by rewriting (3.7) as

$$(R + \alpha_j)\phi_j - \alpha_j\overline{\phi}_j = R\phi_{j-1}, \qquad (3.8)$$

where

$$\phi_j \equiv \left(\frac{\rho_i}{\rho}\right)_j.$$

Successive substitution in (3.8) yields

$$\phi_j\left(1 - \frac{R^4}{R_{1,2,3,4}}\right) = \frac{\alpha_j}{R_j}\overline{\phi}_j + \frac{R\alpha_{j-1}\overline{\phi}_{j-1}}{R_{j,j-1}}$$
$$+ \frac{R^2\alpha_{j-2}\overline{\phi}_{j-2}}{R_{j,j-1,j-2}} + \frac{R^3\alpha_{j-3}\overline{\phi}_{j-3}}{R_{1,2,3,4}}, \qquad (3.9)$$

where

$$\begin{aligned}
R &\equiv \frac{V}{L} \\
R_j &\equiv (R + \alpha_j) \\
R_{j,k} &\equiv (R + \alpha_j)(R + \alpha_k) \\
R_{j,k,l} &\equiv (R + \alpha_j)(R + \alpha_k)(R + \alpha_l) \\
R_{1,2,3,4} &\equiv (R + \alpha_1)(R + \alpha_2)(R + \alpha_3)(R + \alpha_4).
\end{aligned}$$

The reader should attempt to interpret (3.9). For example, ϕ_j clearly depends on the value of $\overline{\phi}$ in each of the boxes, weighted by measures of transport efficiency.

A particularly interesting solution exists in the following limit:

$$\alpha_1 \gg R$$
$$\alpha_2, \alpha_3, \alpha_4 \ll R.$$

Then

$$R_1 \approx +\alpha_1$$
$$R_j \approx R \text{ for } j \neq 1$$

and from (3.9)

$$\phi_1 \approx \overline{\phi_1}$$
$$\phi_2 \approx \overline{\phi_1}$$
$$\phi_3 \approx \overline{\phi_1}$$
$$\phi_4 \approx \overline{\phi_1}.$$

Note, that in this limit, the answer does *not* depend on the sign of M (or V). Also, in the event that $\overline{\phi_1} \gg \overline{\phi_2}, \overline{\phi_3}$, the column density of 'ozone' below boxes 2 and 3 has been greatly increased by transport (beyond what would be implied by photochemical equilibrium). (What would one have to do to make the column density below boxes 2 and 3 greater than it is below boxes 1 and 4? How might this relate to Figure 3.4?)

Exercise

3.1 In terms of our 4-box model, take $p_1 = 52\text{mb}$, $p_2 = 27\text{mb}$, $p_3 = 2\text{mb}$, and $L = 4000\text{km}$. Let $V = 1\text{m/s}$. From Figure 3.1, reasonable winter choices for $\left(\frac{\overline{\rho_i}}{\rho}\right)_j$ are

$$\left(\frac{\overline{\rho_i}}{\rho}\right)_1 = 8\text{ppm}$$

$$\left(\frac{\overline{\rho_i}}{\rho}\right)_2 = 3\text{ppm}$$

$$\left(\frac{\overline{\rho_i}}{\rho}\right)_3 = 1\text{ppm}$$

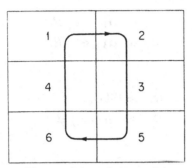

Figure 3.7: 6-box model.

$$\left(\frac{\overline{\rho_i}}{\rho}\right)_4 = 3.5\text{ppm}$$

and from Figure 3.3 reasonable choices for α_j are

$$\alpha_1 = 7\ 10^{-7}\ \text{sec}^{-1}$$
$$\alpha_2 = 2\ 10^{-7}\ \text{sec}^{-1}$$
$$\alpha_3 = 1.5\ 10^{-8}\ \text{sec}^{-1}$$
$$\alpha_4 = 2.5\ 10^{-8}\ \text{sec}^{-1}.$$

Evaluate ϕ_j (for $j = 1, 2, 3, 4$), and compare *roughly* with Figure 3.2. Also evaluate 'column' densities. Discuss differences from the results in Figure 3.4. What happens to your answers when the sign of the large-scale flow is reversed?

If, instead of our 4-box model, we had a 6-box model of the sort shown in Figure 3.7, where boxes 1–4 are as before, while box 5 is characterized by a very long photochemical rate constant, and box 6 is characterized by a very low equilibrium concentration and a very large rate constant, what would you expect to happen to concentrations in boxes 1–4, and to the column densities?

Chapter 4

'Statics' of a rotating system

Supplemental reading:

Holton (1979), chapter 1 and section 3.2

Houghton (1977), sections 7.3–7.6

4.1 Geostrophy and hydrostaticity

In this chapter some terms are introduced which are needed for the subsequent review of observations. Specifically, we will review hydrostaticity and introduce geostrophy – both involve static balances[1]. Our approach will be 'quick and dirty'. Matters will be approached more carefully in Chapter 6. In static balances, forces are balanced by pressure gradients: $-\nabla p$. Consider a rotating fluid in a gravitational field (oriented parallel to the rotation vector; *viz.* Figure 4.1). Let the flow consist simply in rotation about the z-axis: $\Omega + \omega(r)$, where

---

[1] It is expected that most readers will be more or less familiar with the contents of this chapter. It is included for the few who may have need of it. Nevertheless, the material is essential. The reader is urged to make sure that all the exercises that follow this chapter can be comfortably handled.

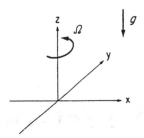

Figure 4.1: Orientation of rotation and gravity in cartesian frame.

$r = \sqrt{x^2 + y^2}$. Moreover, let the flow be steady. The force balance in the z-direction will simply be

$$\frac{\partial p}{\partial z} = -\rho g. \tag{4.1}$$

This balance is referred to as the *hydrostatic balance*. In the r-direction the radial pressure gradient must balance the centrifugal force:

$$\rho(\Omega + \omega)^2 r = \frac{\partial p}{\partial r} \tag{4.2}$$

or

$$\rho(\Omega^2 + 2\Omega\omega + \omega^2)r = \frac{\partial p}{\partial r}.$$

Let

$$p = p_0 + p',$$

where the following expression defines p_0

$$\rho\Omega^2 r = \frac{\partial p_0}{\partial r}$$

($\Omega^2 r$ serves to modify $\vec{g}$ in geophysical systems). So

$$\rho(2\Omega\omega + \omega^2)r = \frac{\partial p'}{\partial r}. \tag{4.3}$$

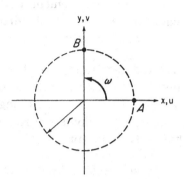

Figure 4.2: Cartesian view of circular motion.

In general[2],

$$\omega \ll \Omega$$

and

$$2\Omega\omega r \cong \frac{1}{\rho}\frac{\partial p'}{\partial r} \tag{4.4}$$

The left-hand side of (4.4) is the Coriolis force per unit mass. This force is merely the linearization of the centrifugal force in a rotating coordinate frame. Equation 4.4 represents what is called *geostrophic* balance, while (4.3) is referred to as *cyclostrophic* balance.

It is sometimes convenient to express (4.4) in cartesian coordinates. Figure 4.2 permits us to view the motion in the horizontal plane within the rotating system. At point A, $dr = dx$, and $\omega r = v$. Equation 4.4 becomes

$$2\Omega v = \frac{1}{\rho}\frac{\partial p}{\partial x}, \tag{4.5}$$

while at point B, $dr = dy$ and $\omega r = -u$, yielding

$$2\Omega u = -\frac{1}{\rho}\frac{\partial p}{\partial y}. \tag{4.6}$$

---

[2]N.B. The quantity $\omega/2\Omega$ is known as the *Rossby number*. More generally, it represents the relative importance of nonlinear inertial terms and the Coriolis force (based on the *system* rotation).

As already mentioned, a more systematic presentation of the above will be given in Chapter 6. The supplemental reading may also be useful to those approaching these topics for the first time.

The following remarks should be kept in mind:

1. Both hydrostaticity and geostrophy are static balances. It is, of course, slightly peculiar to refer to a force balance involving a moving fluid as 'static'. Nevertheless, as long as one refers to a centrifugal force, then geostrophy is a statement that two forces balance rather than a dynamic (prognostic) statement.

2. Neither hydrostaticity nor geostrophy is a causal relation.

3. Strictly speaking, Equation 4.1 is true only in the absence of vertical acceleration (relative to the already rotating system) and friction. Equation 4.4 is always approximate. Nevertheless, for horizontal accelerations with time scales longer than a pendulum day (π/Ω) and vertical accelerations with time scales longer than the Brunt-Vaisala period[3] $(O(5 \text{ minutes}))$, (4.1) and (4.4) remain very nearly true. These conditions apply, for the most part, to the large-scale motions of the atmosphere and oceans (at least away from frictional layers at boundaries).

4. For motions in the atmosphere and oceans the appropriate choice for Ω is its vertical component. The situation is illustrated in Figure 4.3. In (4.4), 2Ω is replaced by $2\Omega \sin \phi$. The quantity $2\Omega \sin \phi$ is generally given the symbol f and is known as the Coriolis parameter.

5. When (4.5) and (4.6) apply, $\nabla \cdot \rho \vec{v} = 0$ (Why?). Also, pressure contours on a horizontal surface are streamlines. That is to say, geostrophic flow is *along* rather than across pressure contours. In this connection, the reader should confirm the *Law of Buy-Ballot*, namely, (in the Northern Hemisphere) when one faces in the direction of geostrophic flow, high pressure is on one's right.

---

[3]This is a quantity which will be defined in Chapter 6. It is a measure of stratification.

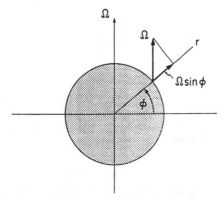

Figure 4.3: The vertical component of the rotation vector on a sphere.

The following sections contain some additional aspects of balanced flow (flow which satisfies (4.1), (4.5), and (4.6)) which will be used in Chapter 5 (on observations).

4.2   Scale height and thickness

In addition to the hydrostatic pressure relation (Equation 4.1) we have the gas law

$$p = \rho RT. \tag{4.7}$$

Using (4.7), (4.1) becomes

$$\frac{\partial p}{\partial z} = -\frac{pg}{RT} \tag{4.8}$$

or

$$\frac{\partial \ln p}{\partial z} = -\frac{1}{H} = -\frac{g}{RT}, \tag{4.9}$$

where

$$H \equiv \frac{RT}{g} = \text{local scale height}$$

$$p = p(\text{surface})e^{-x} \tag{4.10}$$

and

$$x \equiv \int_0^z \frac{dz}{H}.$$

The quantity

$$z^* = \underbrace{H_0}_{characteristic\ scale\ height} \ln\left(\frac{p_{\text{surface}}}{p}\right) \qquad (4.11)$$

is almost like height if H doesn't vary 'too much'. It is also the basis of a popular coordinate system (log–p coordinates). (N.B. For a system in geostrophic and hydrostatic balance, the p-field completely determines the temperature and horizontal wind fields through Equations 4.9, 4.5, and 4.6.)

The hydrostatic relation can be rewritten

$$dp = -\rho g\, dz$$

or

$$d\ln p = -\frac{g}{RT}dz,$$

yielding

$$\int_{z_1}^{z_2} dz \equiv \text{thickness} = -\int_{p_1}^{p_2} \frac{R}{g}T\, d\ln p \qquad (4.12)$$

('thickness' refers to the vertical separation of two isobaric surfaces). The related quantity

$$\Phi \equiv \int_0^z g\, dz = -\int_{p_0}^{p} RT\, d\ln p \qquad (4.13)$$

is referred to as geopotential height. The quantity Φ/g is called the height field.

4.3   Thermal wind and pressure coordinates

It is easily shown[4] that

$$\frac{1}{\rho}\left(\frac{\partial p}{\partial x}\right)_z = \left(\frac{\partial \Phi}{\partial x}\right)_p \tag{4.14}$$

$$\frac{1}{\rho}\left(\frac{\partial p}{\partial y}\right)_z = \left(\frac{\partial \Phi}{\partial y}\right)_p. \tag{4.15}$$

Equations 4.5 and 4.6 then become

$$fu = -\frac{\partial \Phi}{\partial y} \tag{4.16}$$

$$fv = +\frac{\partial \Phi}{\partial x}. \tag{4.17}$$

Now from (4.13), $\frac{\partial \Phi}{\partial p} = -\frac{RT}{p}$. So differentiating (4.16) and (4.17) with respect to p yields

$$p\frac{\partial v}{\partial p} = -\frac{R}{f}\left(\frac{\partial T}{\partial x}\right)_p = \frac{\partial v}{\partial \ln p} \tag{4.18}$$

$$p\frac{\partial u}{\partial p} = +\frac{R}{f}\left(\frac{\partial T}{\partial y}\right)_p = \frac{\partial u}{\partial \ln p}. \tag{4.19}$$

4

$$p = p(x, y, z, t)$$

and

$$dp = \frac{\partial p}{\partial x}dx + \frac{\partial p}{\partial y}dy + \frac{\partial p}{\partial z}dz + \frac{\partial p}{\partial t}dt = 0,$$

which implies

$$\frac{\partial z}{\partial x} = -\left(\frac{\partial p}{\partial x}\right)_z \bigg/ \left(\frac{\partial p}{\partial z}\right)_x = \frac{1}{\rho g}\frac{\partial p}{\partial x},$$

which in turn implies

$$\left(\frac{\partial \Phi}{\partial x}\right)_p = \frac{1}{\rho}\left(\frac{\partial p}{\partial x}\right)_z.$$

Equations 4.18 and 4.19 may be rewritten to yield the *thermal wind* relations (in log–p coordinates):

$$\frac{1}{H}\frac{\partial u}{\partial \ln p} = -\frac{H_0}{H}\frac{\partial u}{\partial z^*} = +\frac{g}{fT}\left(\frac{\partial T}{\partial y}\right)_p \tag{4.20}$$

$$\frac{1}{H}\frac{\partial v}{\partial \ln p} = -\frac{H_0}{H}\frac{\partial v}{\partial z^*} = -\frac{g}{fT}\left(\frac{\partial T}{\partial x}\right)_p. \tag{4.21}$$

Under many circumstances $\frac{H_0}{H} \sim 1$, $z^* \sim z$, and pressure surfaces are far more horizontal than temperature surfaces. Then

$$\frac{\partial u}{\partial z} \approx -\frac{g}{fT}\frac{\partial T}{\partial y} \tag{4.22}$$

$$\frac{\partial v}{\partial z} \approx +\frac{g}{fT}\frac{\partial T}{\partial x}. \tag{4.23}$$

From the above expressions for the *thermal wind* we see that the counterpart of the Law of Buy-Ballot for shear is simply that if one is facing in the direction in which wind is increasing upward, high temperature is on one's right (again for the Northern Hemisphere).

Exercises

Problems on geostrophy and hydrostaticity are pretty much standard in any text on dynamic meteorology. Exercises 4.1–4.6 here are from Houghton (1977); Exercise 4.7 is from Holton (1979).

4.1 What is the pressure gradient required at the earth's surface at 45° latitude to maintain a geostrophic wind velocity of 30 ms^{-1}? What does this suggest about typical pressure gradients?

4.2 The atmospheric surface pressure at radius r_0 from the center of a tornado rotating with constant angular velocity ω is p_0. Show that the surface pressure at the center of the tornado is

$$p_0 \exp\left(-\frac{\omega^2 r_0^2}{2RT}\right)$$

when the temperature is assumed constant.

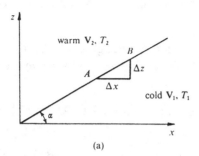

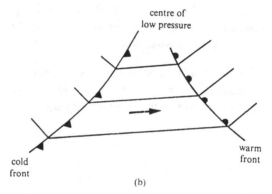

Figure 4.4: (a) Cross section through a front. (b) Warm sector of a typical depression showing warm and cold fronts. From Houghton (1977).

4.3 Figure 4.4a is a cross section of part of a front. At a certain level, the air possesses temperatures T_2 and T_1 ($T_2 > T_1$), respectively, on the two sides of the front, which has a slope of α relative to the horizontal. Apply the hydrostatic equation and the geostrophic wind equation to the region AB, together with the condition that the pressure must be continuous across the frontal surface. Show that in equilibrium the components of velocity v_2 and v_1 in the y-direction on the two sides of the front satisfy the relation

$$(T_1 - T_2)g\tan\alpha = (v_2 T_1 - v_1 T_2)f.$$

If $T_2 - T_1 = 3\text{K}$ and $v_1 - v_2 = 10\text{ms}^{-1}$, find α. The slopes of typical fronts vary from around 1/50 to 1/300. Note that if $f = 0$, that is, the earth were not rotating, the sloping surface could not be in equilibrium.

4.4 From the sense of the velocity change given by the above equation, show that the kink in the isobars is always cyclonic (*viz.* Figure 4.4b).

4.5 The wind at the surface is from the west. At cloud level it is from the south. Do you expect the temperature to rise or fall?

4.6 If the pole is 40K colder than the equator and the surface wind is zero, what wind would you expect at the 200mb pressure level? Compare the temperature and wind fields in Chapter 5, and show that they are *roughly* consistent with your result.

4.7 The planet Venus rotates about its axis so slowly that to a reasonable approximation the Coriolis parameter may be set equal to zero. For steady, frictionless motion parallel to latitude circles Equation 4.3 reduces to

$$\frac{u^2 \tan \phi}{a} = -\frac{1}{\rho} \frac{\partial p}{\partial y}.$$

By transforming this expression to isobaric coordinates show that the *thermal wind* equation in this case can be expressed in the form

$$\omega_r^2(p_1) - \omega_r^2(p_0) = \frac{-R \ln(p_0/p_1)}{(a \sin \phi \, \cos \phi)} \frac{\partial < T >}{\partial y},$$

where R is the gas constant, a is the radius of the planet, and $\omega_r \equiv u/(a \cos \phi)$ is the relative angular velocity. How must $<T>$ (the vertically averaged temperature) vary with respect to latitude in order that ω_r be a function only of height? If the zonal velocity at $\simeq 60$km height above the equator ($p_1 = 2.9 \times 10^5$ Pa) is 100ms^{-1} and the zonal velocity vanishes at the surface of the planet ($p_0 = 9.5 \times 10^6$ Pa), what is the vertically averaged temperature difference between the equator and pole assuming that ω_r depends only on height? The planetary radius is $a = 6100$ km, and the gas constant is $R = 187$J kg^{-1} K^{-1}.

Chapter 5

Observed atmospheric structures

Supplemental reading:

Lorenz (1967)

Palmén and Newton (1967)

Charney (1973)

5.1 General remarks

Our introduction to the observed state of motion and temperature in the atmosphere will be restricted (for the most part) to fairly gross features. Almost no mention will be made of the numerous features most closely associated with the most common perceptions of weather: hurricanes, fronts, thunderstorms, tornadoes, and clear air turbulence, to name a few. While there is something perhaps paradoxical and certainly regrettable about these omissions, the amount of detail required to cover them would far exceed both our time and our capacity for absorption of information. There is an additional reason for restricting ourselves to larger (synoptic) scales: namely, the conventional upper air data network does not resolve the smaller scales.

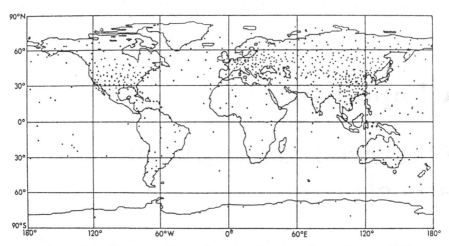

Figure 5.1: Radiosonde station distribution (from Oort, 1978).

The question of resolution is not a simple one and before proceeding, a few remarks on the nature of meteorological data are in order.
Data, in the sense used by experimental sciences (namely, single measurements of an isolated system), are inappropriate to meteorology.
Temporal and spatial variability are such inextricable features of meteorological phenomena that isolated measurements at a single location
are at best inadequate – and usually useless. Figure 5.1 shows the distribution of radiosonde stations at which conventional meteorological
balloon soundings are taken at least once daily (and frequently twice
daily at $0000Z$ and $1200Z$). These soundings consist in pressure, temperature, and humidity measurements transmitted by small radio trans-
mitters to the ground. In addition, the balloons are tracked by radar in
order to obtain profiles of horizontal wind. In principle, such data are
available with fairly high vertical resolution ($O(1\,\mathrm{km})$), but usually the
archived data sets list data only for a subset of the standard levels (surface, 1000mb, 850mb, 700mb, 500mb, 400mb, 300mb, 250mb, 200mb,
150mb, 100mb, 50mb, 30mb, 20mb, 10mb)[1]. From Figure 5.1 we see

---

[1]These are supposed to be standard levels. In addition levels at which extrema
occur are supposed to be recorded as *significant levels*. Unfortunately data sets
frequently omit some of these levels.

that the horizontal distribution of stations – especially over the oceans and in the Southern Hemisphere – is inadequate to resolve any but the coarsest of features. The coverage, however, is extremely nonuniform, and over the Northern Hemisphere continents there is often fairly dense coverage. Temporal coverage is barely adequate to resolve five-day periods (one usually needs 4–6 points to resolve a period or wavelength), and height resolution, while adequate for some purposes, is usually inadequate for the resolution of boundary layers commonly observed (in more detailed measurements) below 700mb.

The limitations imposed by the crude nature of our meteorological measurements are substantial – and real. Moreover, the 'raw' data in tabular form (or on tape) is peculiarly uninformative. Indeed, if each station produced an independent uncorrelated time series, it would be difficult to know where to begin any theoretical description (to be sure, we could then begin to formulate statistics, and attempt to explain these statistics), but, fortunately, when the data are presented in the form of global or regional maps, we see that patterns emerge which seem to evolve in a traceable manner. It is, in the form of these maps, that we usually study the data. Such maps are the basis of much of this chapter. A quick glance at these maps shows representations which are continuous over the whole globe (or at least a hemisphere), while the 'raw' data come from a relatively few isolated stations. Clearly, the maps are not exactly data; they include a very substantial amount of interpolation – and this interpolation is rarely as straightforward as simple linear interpolation. Such maps are referred to as *analyzed* data. When the analysis is performed by hand by a synoptic meteorologist, it is referred to as 'subjective' analysis. The contours drawn in data-free regions probably contain useful information – especially when prepared by an experienced meteorologist – since much is known about the expected time and space evolution of disturbances. However, there is no getting away from the fact that what is drawn is *not* data. This becomes particularly disturbing when the contours show substantial detail in data-free regions. In general, these details will differ in different analyses.

When the analysis is performed according to fixed rules and algorithms, the analysis is referred to as 'objective'. The advantage of 'objective' analyses is their reproducibility, but there is no other *a priori* guaranty of greater accuracy than that found in subjective analyses. Recently, it has become common to analyze data with the aid of numerical weather prediction models. Iterative use of predictions over short periods allows one to interpolate the raw data in a manner which is somehow consistent with the model physics. Moreover, such schemes allow the 'assimilation' of data other than the standard radiosonde data obtained from the stations in Figure 5.1. Of primary importance in this regard is the satellite obtained infrared radiance (from which coarse vertical temperature structure may be inferred)[2], and winds obtained from jumbo jets with accurate inertial guidance systems. Such model based analysis-assimilation schemes are impressive. Tests show that the interpolations are frequently surprisingly accurate. There is a general consensus that these new objective analyses are consideraby better than older subjective analyses – especially over the oceans. However, even these analyses are limited by (among other things) the model physics and resolution. In general, such models have only primitive parameterizations of cumulus convection, turbulence, radiative transfer, and sub-grid transfers by gravity waves. The forecast skill of such models is frequently poor in the tropics. Analysis-assimilation schemes which emphasize compatibility with the model physics often throw away actual data in order to produce a compatible analysis. Moreover, analyses based on different models frequently differ substantially (Lau and Oort, 1981).

The above description may be unduly critical. Nevertheless, there can be no doubt that the comparison of theory and data is a more difficult proposition in meteorology than in the traditional sciences. Science consists in a creative tension between theory and data: theory explaining data – data testing theory. In each case, the data consists in numbers with error bars estimating the likely uncertainty of the data. Such error bars can be attached to the results of individual soundings, but no similar methodology is readily attached to analyzed data where

---

[2]The global coverage afforded by satellite radiance measurements would seem likely to greatly improve global forecasts and analyses. This, in fact, has proven to be the case in the Southern Hemisphere, where there is almost no other data. However, in the Northern Hemisphere little or no improvement has been obtained. Apparently, inaccuracies and poor vertical resolution have limited the impact of satellite data when other data are available.

whole weather systems may be missed while occasionally systems are drawn which in reality didn't exist. The situation becomes even more questionable when higher levels of analysis are introduced: that is, spectral decompositions, correlations, etc. While such analyses might be applied, in principle, to the raw data, the methods are far better suited to regularly spaced data. In general, such regularly spaced, 'gridded' data are obtained from analyzed maps. In this chapter, we will avoid such products of higher level analyses – though the results can often be informative and suggestive.

At this point it might appear that analyzed data is completely uncertain. This is certainly untrue – even though the quantitative measures of uncertainty have not been adequately developed. Certain important features on maps appear regularly and clearly (high signal–to–noise ratio) and are readily related to our tangible experience of weather. Analyzed data do, in fact, help us to isolate and quantify such features. However, 'data' as used in meteorology are not quite so concrete as carefully obtained laboratory data – and may, on occasion, even be wrong!

In describing the large–scale structure of the atmosphere, I will assume that the reader is already familiar with the variation of the horizontally averaged temperature with height. Similarly, the reader should be familiar with the typical heights at which various pressures occur. Finally, more detailed examination of the data for specific phenomena will be made in some later chapters.

5.2   Daily and monthly maps

We begin our study with eight figures, each of which consists of a set of four hemispheric maps: one for January 15, 1983, one for July 16, 1983, one for the average over January 1983, and one for the average over July 1983. The maps are all from the European Centre for Medium Range Weather Forecasting (ECMWF). Figures 5.2 and 5.6 show pressure contours at sea level (corrected for topography) for the Northern and Southern Hemispheres, respectively. Figures 5.3–5.5 show height fields for 500mb, 300mb, and 50mb in the Northern Hemisphere while Figures 5.6–5.9 show the same for the Southern Hemisphere.

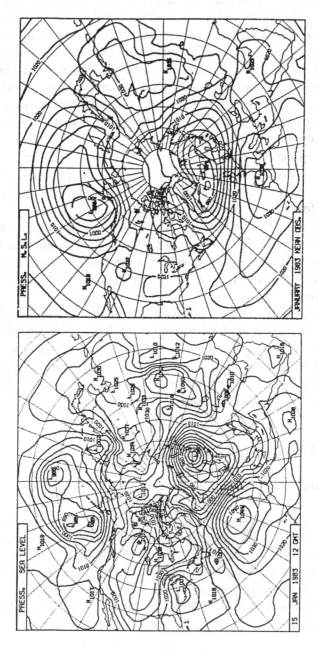

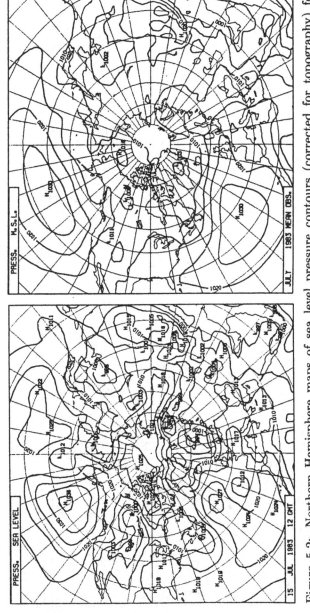

Figure 5.2: Northern Hemisphere maps of sea level pressure contours (corrected for topography) for January 15, 1983 and July 16, 1983 and monthly mean maps for January and July of 1983. (millibars)

43

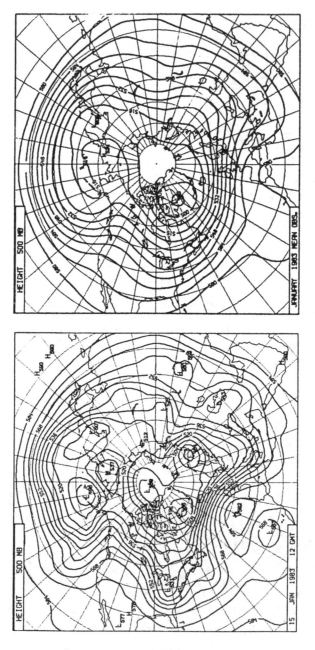

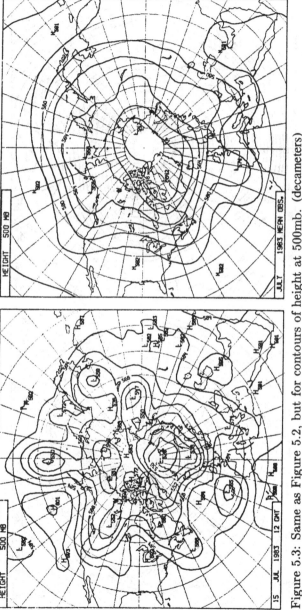

Figure 5.3: Same as Figure 5.2, but for contours of height at 500mb. (decameters)

45

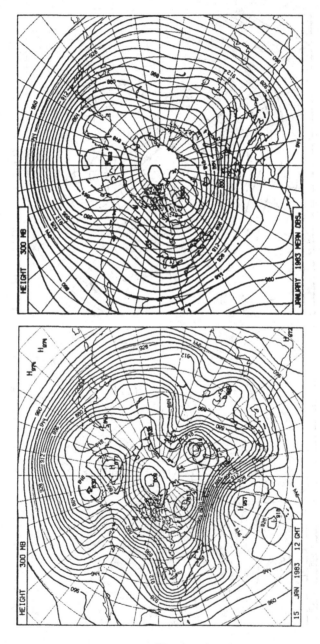

46

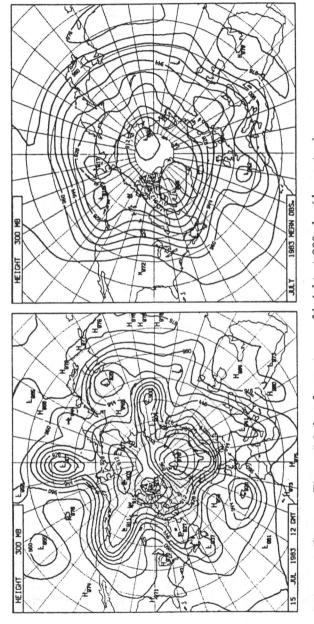

Figure 5.4: Same as Figure 5.2, but for contours of height at 300mb. (decameters)

47

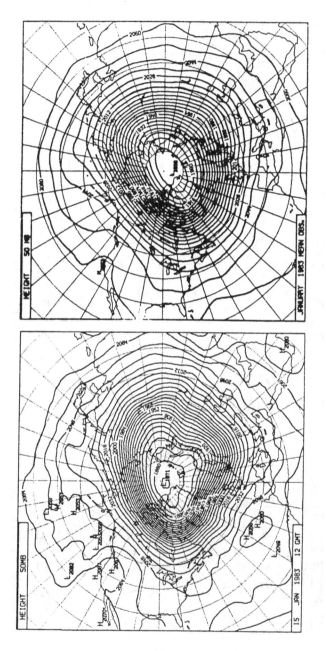

48

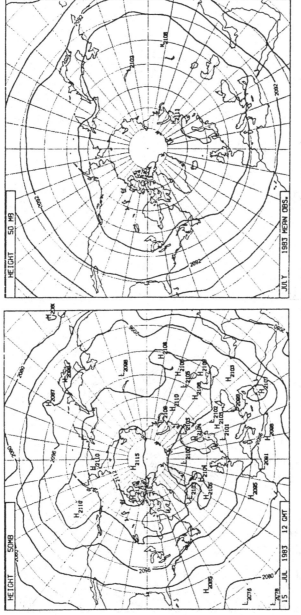

Figure 5.5: Same as Figure 5.2, but for contours of height at 50mb. (decameters)

49

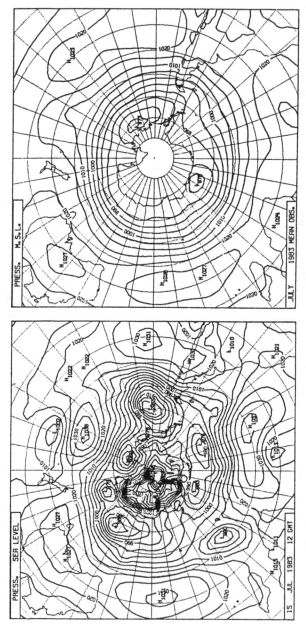

50

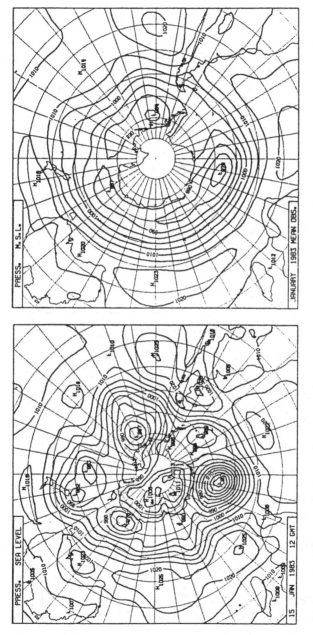

Figure 5.6: Same as Figure 5.2, but for Southern Hemisphere. (millibars)

51

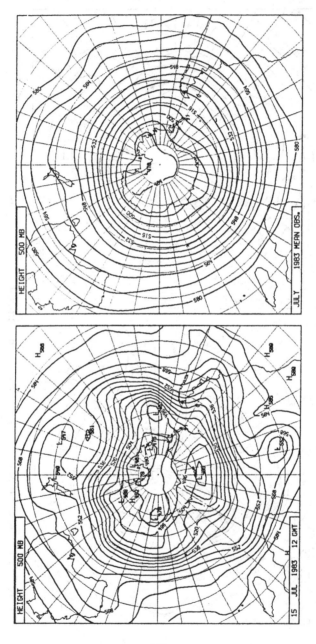

52

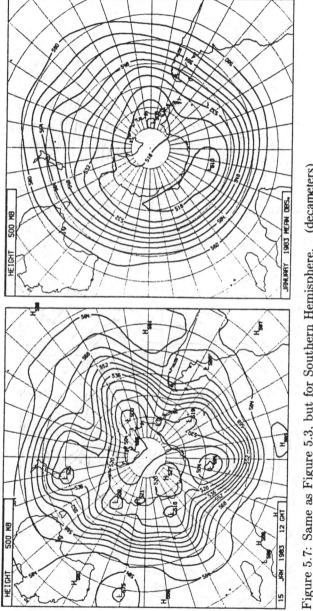

Figure 5.7: Same as Figure 5.3, but for Southern Hemisphere.   (decameters)

53

54

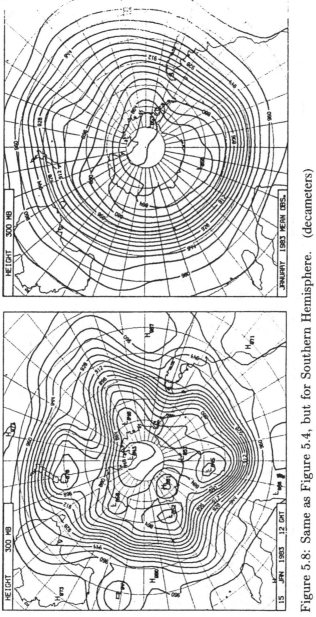

Figure 5.8: Same as Figure 5.4, but for Southern Hemisphere. (decameters)

55

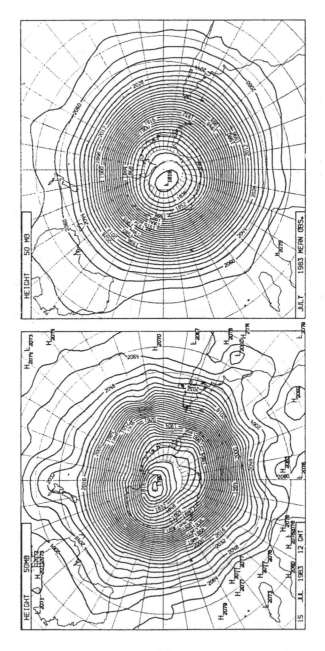

56

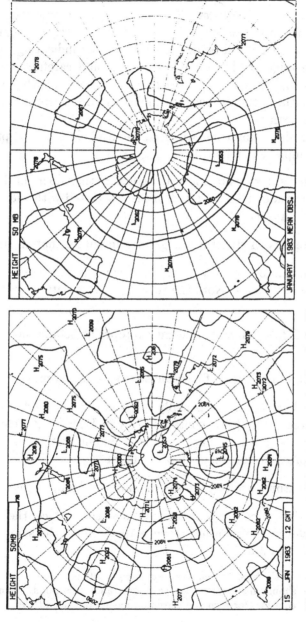

Figure 5.9: Same as Figure 5.5, but for Southern Hemisphere. (decameters)

57

Such maps display many phenomena – though the maps presented here are insufficient to adequately isolate and quantitatively delineate such phenomena. Nevertheless, the maps warrant close and thoughtful scrutiny. In these notes, we will focus on the Northern Hemisphere maps – and even that will be done briefly, simply indicating the kinds of things one might look for. The reader should carefully study the Southern Hemisphere maps in order to see in what ways the meteorology of the Southern Hemisphere resembles and differs from that of the Northern Hemisphere.

Figure 5.2 shows maps of Northern Hemisphere sea level pressure. Note, for example, the following:

1. Daily maps show more fine-scale structure than do monthly means implying that fine-scale structure is associated with shorter time scales than a month.

2. The intensity of structures is greater in winter than in summer.

3. Note the large-scale lows over the oceans and highs over land in the January mean. Note the reversal of this pattern in the July mean.

In Figure 5.3, which shows height contours at 500mb, note the following:

1. Again note the loss of fine-scale structure in the monthly means.

2. Fine-scale structure (especially for January 15) tends to be less associated with closed contours than in Figure 5.2. This is *not* because these features are weaker at 500mb; rather, it is due to the stronger mean zonal[3] flow at 500mb[4].

3. Again features are less intense in summer than in winter.

4. Notice that in contrast to the results at the surface, the phase of the waves in the monthly mean maps is much the same in both January and July.

---

[3] *Zonal* refers to the west-to-east direction along a latitude circle.

[4] The reader should make sure that he understands this point. If necessary, synthesize some contours for assumed wave and mean flow magnitudes.

Turning to Figure 5.4, which shows height contours at 300mb, we see a rather substantial similarity to Figure 5.3 except for a general intensification of the zonal mean flow and the eddies (deviations from the zonal mean).

In moving to Figure 5.5, which shows height contours at 50mb, from Figure 5.4, we are moving across the tropopause. We now see a close similarity between the daily maps and the monthly means indicating a relative absence of short-period features. We also see an increase in the dominant spatial scale. In the summer we see a pronounced reduction in both eddies and the zonal mean.

The reader can confirm from Figures 5.6-5.9 the presence of most of the above features in the Southern Hemisphere. However, the monthly means show much less wave structure – presumably due to the relative absence of land in the Southern Hemisphere. Nevertheless, what wave structure there is in the monthly means for the Southern Hemisphere is much the same in both amplitude and phase for both January and July. In the Northern Hemisphere, the monthly mean waves were significantly stronger in January.

5.3 Zonal means

5.3.1 Seasonal means

As is evident from the preceding figures, maps display the superposition of many phenomena and systems. As such, they are difficult to analyze unambiguously. It is usual to process the maps in such a manner as to isolate some subset of what is going on. Taking monthly means is an example of such processing. Another example, of great historical importance in meteorology, is the taking of zonal means in order to study the height and latitude variations of such means. Figures 5.10 and 5.11 show the meridional sections of zonally averaged zonal wind and temperature for each season (from Newell *et al.*, 1972).

The following are a few of the features which may be noted in these figures:

1. Regardless of season, surface winds tend to be easterly (from the east) within 30° of the equator and westerly poleward of 30°.

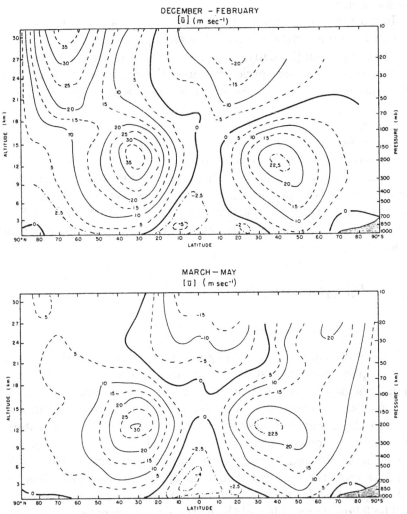

Figure 5.10: Meridional section of zonally averaged zonal wind for each of the four seasons (from Newell *et al.*, 1972).

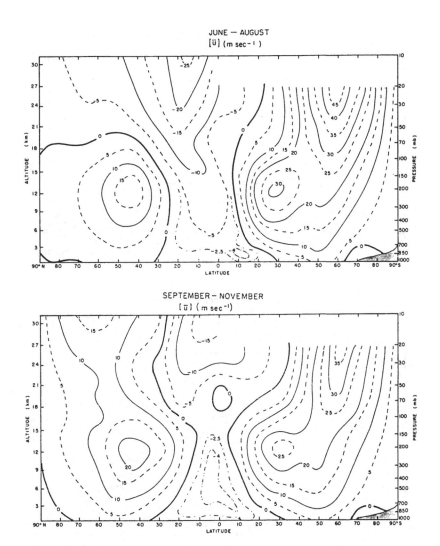

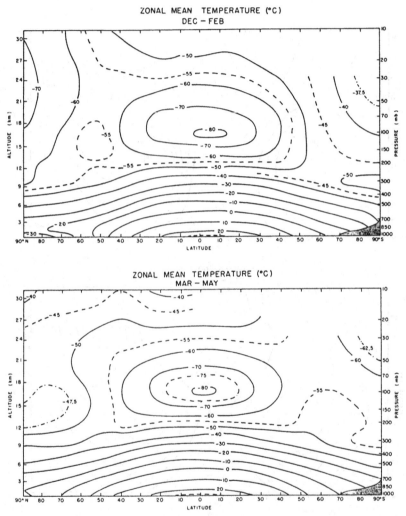

Figure 5.11: Same as Figure 5.10, but for zonally averaged temperature.

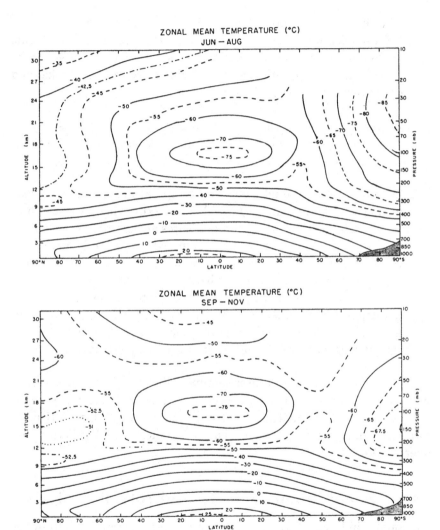

ZONAL MEAN TEMPERATURE (°C)
JUN – AUG

ZONAL MEAN TEMPERATURE (°C)
SEP – NOV

Surface easterlies tend to be stronger on the winter side of the equator.

2. The midlatitude troposphere is characterized by westerly jets in both hemispheres. The jet maxima occur at about 12km altitude ($\sim$ 200mb). The winter maxima are stronger ($\sim 30 - -50$m/s) and occur near 30° latitude. The summer maxima are weaker and occur further poleward ($\sim$ 45° latitude).

3. In the winter stratosphere, there is also a polar night westerly jet centered near 60°. This jet is generally stronger in the Southern Hemisphere winter.

4. Zonally averaged zonal winds and temperatures are pretty nearly in thermal wind balance. Thus, below 12km, where westerly flow is increasing with altitude, temperatures are decreasing away from the equator. However, above 12km, where westerly winds are decreasing with height, we have minimum temperatures at the equator. In the summer hemisphere, temperatures increase monotonically to the pole. However, in the winter hemisphere, a temperature maximum is reached in the lower stratosphere near 50° and the temperature falls rapidly poleward of this latitude (consistent with the presence of the polar night jet).

Figure 5.12 shows the winter–summer differences in zonally averaged temperature (taken from Newell *et al.*, 1972). Not surprisingly, these differences are small in the tropics and large at high latitudes. Notice as well the differences between the Northern and Southern Hemispheres (Why?).

5.3.2   Zonal inhomogeneity and rôle of analysis

The zonal wind at any given longitude will, of course, differ from the zonal average. December–February sections for various longitudes are shown in Figure 5.13; June–August sections are shown in Figure 5.14. Inhomogeneity is greatest in the Northern Hemisphere where values about double the zonal average may be found at some longitudes. This

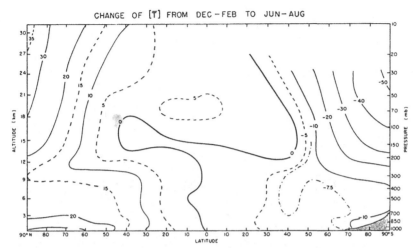

Figure 5.12: Winter–summer differences in zonally averaged temperature (from Newell *et al.*, 1972).

is also seen in Figure 5.15, where Northern Hemisphere winter contours of zonal wind at 200mb are shown. Also, as already noted in the beginning of this chapter, the data we are dealing with are analyzed data. Figure 5.15 shows the results of two different analyses (GFDL and NMC) and the differences between the two analyses. The differences are on the order of 10–20%, which is a plausible measure of our uncertainty.

5.3.3 Middle atmosphere

So far we have concentrated our attention on the zonally averaged zonal wind and temperature in the troposphere and lower stratosphere. Figure 5.16 shows these quantities up to the lower thermosphere. The data is primarily Northern Hemisphere data; the right and left halves of the sections correspond to winter and summer. Note the monsoonal nature of the mesospheric winds: winter is characterized by westerlies; summer by easterlies. Note also that temperature increases monotonically from the winter pole to the summer pole at the stratopause (*ca.* 50km), but

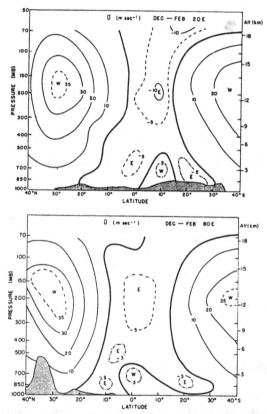

Figure 5.13: Meridional sections of December–February zonal winds at specific longitudes (From Newell *et al.*, 1972)

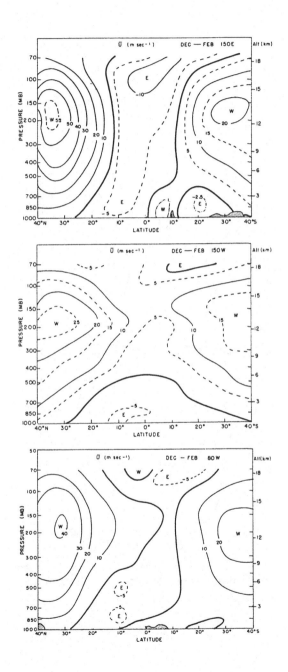

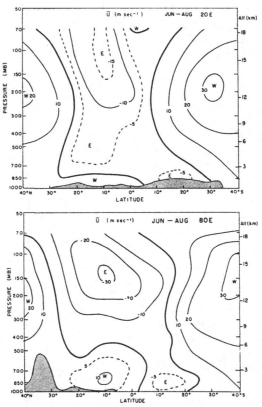

Figure 5.14: Meridional sections of June–August zonal winds at specific longitudes (from Newell *et al.*, 1972)

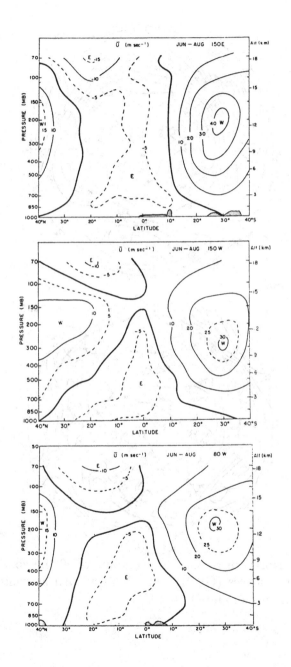

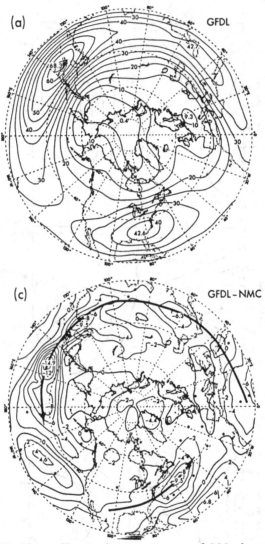

Figure 5.15: Northern Hemisphere contours of 200mb zonal winds for two different analyses (GFDL and NMC) as well as contours of the differences (from Lau and Oort, 1981).

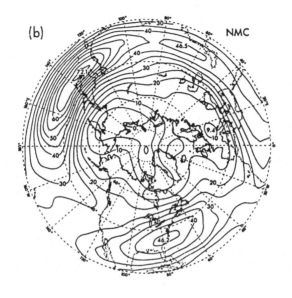

$\bar{U}$ 200mb
WINTER
ms^{-1}

at the mesopause (*ca.* 83km) the temperature is increasing monotonically from the summer pole to the winter pole. This is again consistent with thermal wind balance.

5.3.4   Quasi-biennial and semiannual oscillations

Figure 5.16 assumes that stratospheric and mesospheric winds are dominated by an annual (12 month) cycle. Near the equator this turns out to be untrue. Figure 5.17 shows a time-height section of zonal wind at Canton Island (2°46'S) (which turns out to be characteristic of the zonal average). Note that zonal winds form a downward propagating wave-like structure with a period of about 26 months; this is referred to as the *quasi-biennial oscillation* which dominates the tropical zonal wind between 16km and 30km. The situation up to 56km is shown in Figure 5.18. Here we see that above about 32km, a *semiannual oscillation* is dominant. Figures 5.17 and 5.18 are based on monthly means. It turns out that stratospheric tropical zonal winds also sometimes display important short period oscillations. In Figure 5.19 we see an oscillation with a period of about 12 days. This has been identified as an *equatorial Kelvin wave*[5].

5.3.5   Stratospheric sudden warmings

Figures 5.5 and 5.16 suggest a rather regular seasonal behaviour in the middle and high latitude stratosphere. However, occasionally (every 1–4 years or so) this pattern breaks down rather spectacularly in a midwinter *stratospheric sudden warming* where the winter pattern breaks down and a summer pattern onsets – during the winter polar night. Figure 5.20 shows a Northern Hemisphere map for January 25, 1957 – just prior to the onset of such a warming. The pattern is almost identical to that in Figure 5.5. (Note that North America is at the bottom of Figure 5.20, while it is at the left in Figure 5.5; note also that height in Figure 5.20 is in 100s of feet while in Figure 5.5 it is in tens of meters.) However, by February 4, 1957, Figure 5.21 shows a significant change: zonal wavenumber two has amplified strongly and

---

[5]We will explain what this is in Chapter 11.

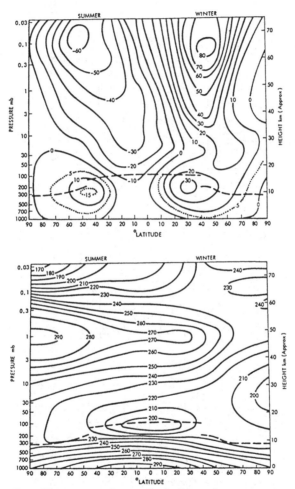

Figure 5.16: Zonally averaged zonal winds and temperatures up to the lower thermosphere (from Andrews, Holton, and Leovy, 1987).

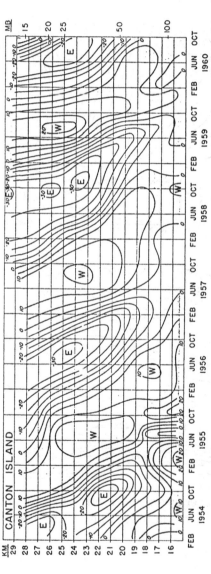

Figure 5.17: Time-height section of monthly mean stratospheric zonal winds at Canton Island (2°46'S) (from Reed and Rogers, 1962).

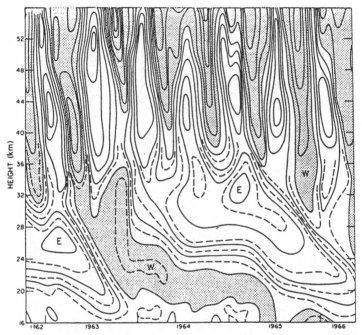

Figure 5.18: Time-height section of near-equatorial monthly mean zonal winds between 16km and 56km (from Wallace, 1973).

The solid contour intervals are 10ms^{-1}. Shaded regions refer to westerlies, unshaded to easterlies.

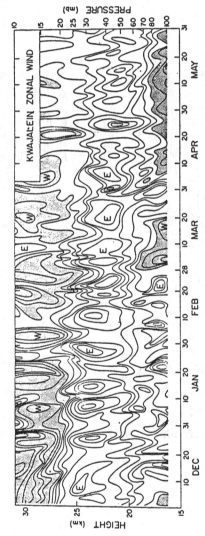

Figure 5.19: Time-height section of daily zonal winds over Kwajalein (from Wallace and Kousky, 1968). The contour intervals are 5ms⁻¹.

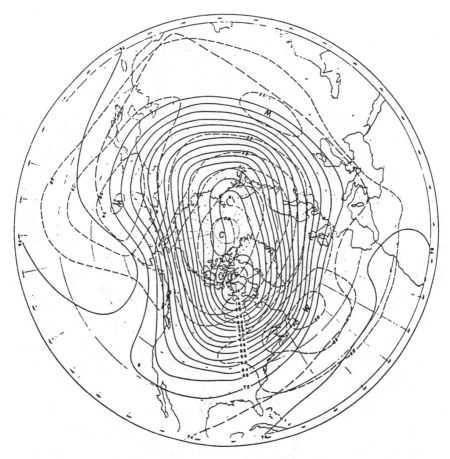

Figure 5.20: Northern Hemisphere 50mb height map for January 25, 1957 (from Reed, Wolfe, and Nishimoto, 1963).

The solid contours refer to height in 100s of feet. The dashed contours refer to temperature in units of °C.

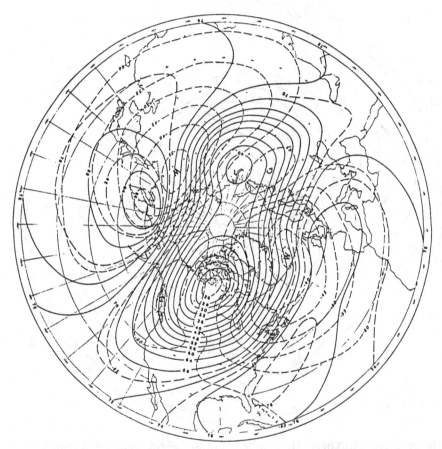

Figure 5.21: Same as Figure 5.20, but for February 4, 1957.

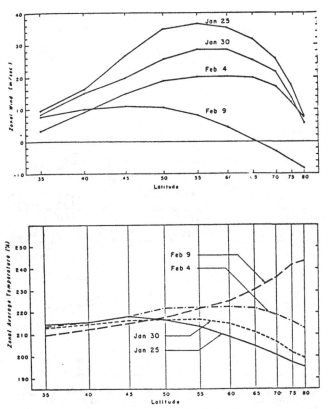

Figure 5.22: Zonally averaged zonal winds and temperatures at 50mb as functions of latitude for various times in the course of a sudden warming (from Reed *et al.*, 1963).

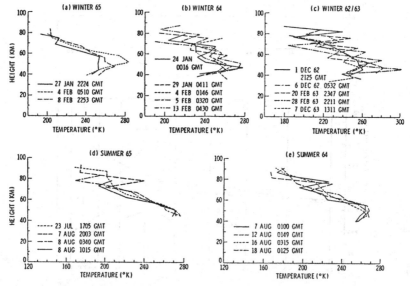

Figure 5.23: Rocket soundings of temperature over Wallops Island (38°N) (from Theon *et al.*, 1967).

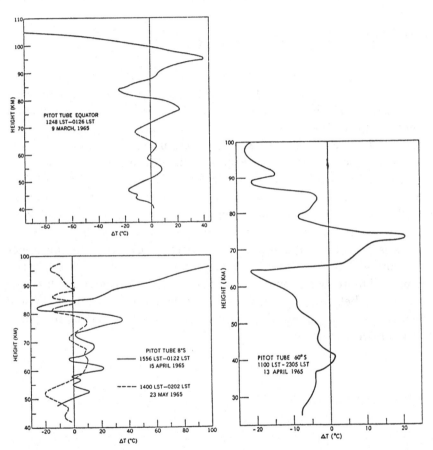

Figure 5.24: Day–night temperature difference profiles obtained from rockets (from Theon *et al.*, 1967).

seems to have drifted westward. This change is accompanied by a zonally averaged poleward heat flux which leads to the zonally averaged changes in zonal flow and temperature shown in Figure 5.22. Observe the change of the temperature minimum at the pole into a temperature maximum. Note also the change from Arctic westerlies to easterlies.

5.4   Short period phenomena

Finally we return briefly to our initial comment on the omission of shorter scale phenomena. It is increasingly recognized that some of these phenomena, in the form of vertically propagating waves, play an essential rôle in the large scale circulation. This will be discussed in later chapters. Such waves have been long noticed in rocket data, examples of which are shown in Figure 5.23. Notice the large amplitude perturbations ($\sim 20°C$) with vertical wavelengths of from 10–15km. These perturbations tend to become obvious at lower altitudes in winter. Rocket data for day–night temperature differences are also of some interest. Examples are shown in Figure 5.24. What stands out in these figures is the very large day–night differences observed in the upper mesosphere ($O(20°C)$) and the fact that as often as not, the night is warmer. The reader should consider what this implies about how the atmosphere behaves – and, in particular, how the atmosphere 'knows' when it is day or night. The implications are by no means restricted to the upper atmosphere.

Chapter 6

Equations of motion

Supplemental reading:

Holton (1979), chapters 2 and 3 deal with equations, section 2.3 deals with spherical coordinates, section 2.4 deals with scaling, and section 3.1 deals with pressure coordinates.

Houghton (1977), chapter 7 deals with equations, and section 7.1 deals with spherical coordinates.

Serrin (1959)

As has been mentioned in the Introduction, it is expected that almost everyone reading these lecture notes (and despite TEXification, these are only notes) will have already seen a derivation of the equations. I have, therefore, decided to cover the equations using Serrin's somewhat less familiar approach.

6.1   Coordinate systems and conservation

Let $\vec{x} = (x_1, x_2, x_3)$ be a fixed spatial position; this will be referred to as an *Eulerian* coordinate system. Now, at some moment $t = 0$ let's look at a fluid and label each particle of the fluid $\vec{X} = \vec{X}(t, \vec{x}) = (X_1, X_2, X_3)$, where $X_i|_{t=0} = x_i$; that is, we label each particle by its position at $t = 0$; this will be referred to as a *Lagrangian* coordinate system. In general, each coordinate system may, in principle, be transformed into the other:

$$\text{Eulerian:} \qquad \vec{x}, t \quad \vec{x} = \vec{x}(\vec{X}, t)$$

$$\text{Lagrangian:} \qquad \vec{X}, t \quad \vec{X} = \vec{X}(\vec{x}, t)$$

Let the velocity of a fluid 'particle' be $\vec{u} = (u_1, u_2, u_3)$.

$$u_i = \frac{Dx_i}{Dt} = \left(\frac{\partial x_i}{\partial t}\right)_{\vec{X} \text{ constant}}$$

Similarly, let $\vec{a}$ be the acceleration of a fluid 'particle':

$$a_i = \left(\frac{\partial^2 x_i}{\partial t^2}\right)_{\vec{X} \text{ constant}} = \frac{\partial u_i}{\partial t} + \frac{\partial u_i}{\partial x_j}\frac{Dx_j}{Dt},$$

where the *summation convention* is used; that is, we sum over repeated indices.

The laws of physics are fundamentally conservation statements concerning $\frac{D}{Dt}$ of something following the fluid. Let us, for the moment, deal with some unspecified field $f(x_i, t)$ (per unit mass):

$$\frac{Df}{Dt} = \frac{\partial f}{\partial t} + \frac{\partial f}{\partial x_i}\frac{Dx_i}{Dt}.$$

Now consider some region of space $R(\vec{x})$. We wish to evaluate

$$\frac{D}{Dt}\int_{R(\vec{x})} f\rho\, d^3x,$$

where ρ is density. A difficulty arises since the fluid within $R(\vec{x})$ is changing. We deal with this by switching to Lagrangian coordinates:

$$\frac{D}{Dt}\int_{R(\vec{x})} f\rho\, d^3x = \frac{D}{Dt}\int_{R(\vec{X})} f\rho J\, d^3X,$$

where

$$
J = \left| \frac{\partial(x_i)}{\partial(X_i)} \right| =
\begin{vmatrix}
\frac{\partial x_1}{\partial X_1} & \frac{\partial x_1}{\partial X_2} & \frac{\partial x_1}{\partial X_3} \\[6pt]
\frac{\partial x_2}{\partial X_1} & \frac{\partial x_2}{\partial X_2} & \frac{\partial x_2}{\partial X_3} \\[6pt]
\frac{\partial x_3}{\partial X_1} & \frac{\partial x_3}{\partial X_2} & \frac{\partial x_3}{\partial X_3}
\end{vmatrix}.
$$

Since $\vec{X}$ is fixed in the moving fluid,

$$
\frac{D}{Dt} \int_{R(\vec{x})} f\rho \, d^3x = \int_{R(\vec{X})} \frac{D}{Dt}(f\rho J) \, d^3X.
$$

For conservation of mass, we take $f = 1$. Then

$$
\frac{D}{Dt} \int_{R(\vec{x})} \rho \, d^3x = \int_{R(\vec{X})} \frac{D}{Dt}(\rho J) \, d^3X = 0,
$$

and since R is arbitrary,

$$
\frac{D}{Dt}(\rho J) = 0. \tag{6.1}
$$

This is a somewhat peculiar form of the continuity equation. It is, however, easily converted to the usual form:

$$
\frac{D}{Dt}(\rho J) = \rho \frac{DJ}{Dt} + J \frac{D\rho}{Dt},
$$

$$
\begin{aligned}
\frac{1}{J}\frac{DJ}{Dt} &= \frac{\frac{\partial(u_1,x_2,x_3)}{\partial(X_1,X_2,X_3)}}{\frac{\partial(x_1,x_2,x_3)}{\partial(X_1,X_2,X_3)}} + \frac{\frac{\partial(x_1,u_2,x_3)}{\partial(X_1,X_2,X_3)}}{\frac{\partial(x_1,x_2,x_3)}{\partial(X_1,X_2,X_3)}} + \frac{\frac{\partial(x_1,x_2,u_3)}{\partial(X_1,X_2,X_3)}}{\frac{\partial(x_1,x_2,x_3)}{\partial(X_1,X_2,X_3)}} \\[10pt]
&= \frac{\partial u_1}{\partial x_1} + \frac{\partial u_2}{\partial x_2} + \frac{\partial u_3}{\partial x_3} = \nabla \cdot \vec{u}
\end{aligned}
$$

and

$$\frac{D\rho}{Dt} + \rho \nabla \cdot \vec{u} = 0 \tag{6.2}$$

Returning to the general case

$$
\begin{aligned}
\frac{D}{Dt} \int_{R(\vec{x})} f\rho \, d^3x &= \int_{R(\vec{X})} \frac{D}{Dt}(f\rho J) \, d^3X \\
&= \int_{R(\vec{X})} \left\{ \frac{Df}{Dt}\rho J + f \underbrace{\frac{D}{Dt}(\rho J)}_{=0} \right\} d^3X \\
&= \int_{R(\vec{X})} \frac{Df}{Dt}\rho J \, d^3X \\
&= \int_{R(\vec{x})} \frac{Df}{Dt}\rho \, d^3x,
\end{aligned}
$$

that is, we can move $\frac{D}{Dt}$ inside an integral within Eulerian space. Note also, $\frac{D}{Dt}$ is applied to f *not* ρf.

6.2   Newton's second law – for fluids

Newton's second law for a volume of fluid, R, is

$$\underbrace{\frac{D}{Dt} \int_R \rho u_i \, d^3x}_{momentum} = \underbrace{\int_R \rho f_i \, d^3x}_{body\ force}$$

$$+ \underbrace{\int_S \sigma_{ij} n_j \, dS,}_{force\ exerted\ on\ surface\ of\ R\ by\ fluid\ outside\ R} \tag{6.3}^{i}$$

N.B. $\int_S \sigma_{ij} n_j \, dS$ can be rewritten $\int_S F_i \, dS$, where the surface force, $\vec{F}$, is given by $F_i = \sigma_{ij} n_j$ ($\vec{n}$ is the outward normal). The stress tensor, σ_{ij}, represents the flux of i-momentum in the j-direction. Intuitively, we expect the flux of i-momentum in the i-direction to be related to pressure.

Now,

$$\frac{D}{Dt} \int_R \rho u_i \, d^3x = \int_R \rho \frac{Du_i}{Dt} d^3x.$$

Also, by the divergence theorem,

$$\int_S \sigma_{ij} n_j \, dS = \int_R \frac{\partial \sigma_{ij}}{\partial x_j} d^3x.$$

Finally, since R is arbitrary, we have

$$\rho \frac{Du_i}{Dt} = \rho f_i + \frac{\partial \sigma_{ij}}{\partial x_j}. \tag{6.4}$$

Note that ρ is outside the derivative. Equation 6.4 is not the usual form of the momentum equation (in particular, the pressure gradient term is buried in $\frac{\partial \sigma_{ij}}{\partial x_j}$); as our first step in evaluating $\frac{\partial \sigma_{ij}}{\partial x_j}$, we will consider *angular momentum*:

$$\frac{D}{Dt} \int_R \vec{r} \times \rho \vec{u} \, d^3x = \int_R \rho \vec{r} \times \vec{f} \, d^3x + \int_S \vec{r} \times \vec{F} \, dS, \tag{6.5}$$

where $\vec{F} = \sigma_{ij} n_j \hat{i}_i$[1]. (Note that (6.5) assumes no intrinsic torques.)

---

[1]The quantity $\hat{i}_i$ is a unit vector in the i-direction.

Rewriting (6.5),

$$\underbrace{\int_R \rho \frac{D}{Dt}(\vec{r} \times \vec{u})\, d^3x}_{A} = \underbrace{\int_R \rho \vec{r} \times \vec{f}\, d^3x}_{B} + \underbrace{\int_R \hat{\imath}_i \frac{\partial}{\partial x_l}(\epsilon_{ijk} x_j \sigma_{kl})\, d^3x}_{C} \quad ^2.$$

$$A = \int_R \rho \left[\underbrace{\overbrace{\frac{D\vec{r}}{Dt}}^{\vec{u}} \times \vec{u}}_{=0} + \vec{r} \times \frac{D\vec{u}}{Dt} \right] d^3x$$

$$B = \int_R \vec{r} \times \underbrace{\left(\rho \frac{D\vec{u}}{Dt} - \frac{\partial \sigma_{ij}}{\partial x_j} \hat{\imath}_i \right)}_{= \vec{f} \text{ from Newton's second law}} d^3x$$

$$C = \int_R \hat{\imath}_i \left[\epsilon_{ijk} \delta_{jl} \sigma_{kl} + \epsilon_{ijk} x_j \frac{\partial \sigma_{kl}}{\partial x_l} \right] d^3x.$$

After obvious cancellation, we are left with

$$\int_R \hat{\imath}_i \epsilon_{ijk} \delta_{jl} \sigma_{kl}\, d^3x = 0$$

or

$$\hat{\imath}_i \epsilon_{ijk} \delta_{jl} \sigma_{kl} = 0$$

or

$$\hat{\imath}(\sigma_{32} - \sigma_{23}) + \hat{\jmath}(\sigma_{13} - \sigma_{31}) + \hat{k}(\sigma_{12} - \sigma_{21}) = 0.$$

Thus, in the absence of intrinsic torques

$$\sigma_{ij} = \sigma_{ji}. \tag{6.6}$$

---

[2]The quantity ϵ_{ijk} is called an alternant and is defined as follows

$$
\begin{aligned}
\epsilon_{ijk} &\equiv 1 \text{ for } ijk = 123,\ 231,\ 312 \\
&\equiv -1 \text{ for } ijk = 321,\ 213,\ 132 \\
&\equiv 0 \text{ when any two of } ijk \text{ are equal.}
\end{aligned}
$$

6.3 Energy

Let us first look at the rate of change of *mechanical* energy. Take the momentum equation (6.4), multiply by u_i and sum over i:

$$\rho \frac{D}{Dt}\left(\frac{u_i u_i}{2}\right) = \rho f_i u_i + u_i \frac{\partial \sigma_{ij}}{\partial x_j}. \tag{6.7}$$

Integrate (6.7) over region R

$$\overbrace{\frac{D}{Dt}\int_R \frac{1}{2}\rho u_i u_i \, d^3x}^{A} = \overbrace{\int_R \rho f_i u_i \, d^3x}^{B} + \int_R u_i \frac{\partial \sigma_{ij}}{\partial x_j} d^3x.$$

The last term can be rewritten

$$
\begin{aligned}
\int_R u_i \frac{\partial \sigma_{ij}}{\partial x_j} d^3x &= \int_R \frac{\partial}{\partial x_j}(u_i \sigma_{ij})\, d^3x - \int_R \frac{\partial u_i}{\partial x_j} \sigma_{ij}\, d^3x \\
&= \underbrace{\int_S u_i \sigma_{ij} n_j \, dS}_{C} - \underbrace{\int_R \frac{\partial u_i}{\partial x_j}\sigma_{ij}\, d^3x}_{D}.
\end{aligned}
$$

The labelled terms are interpreted as follows:

Term A: Time rate of change of mechanical energy.

Term B: Work done by body forces.

Term C: Work done by surface stresses.

Term D: Needs elucidation!

Recall, there is no conservation of mechanical energy alone. What about total energy?

$$
\begin{aligned}
\frac{D}{Dt}\int_R \rho\left(\frac{u^2}{2}+e\right) d^3x &= \int_R \rho f_i u_i \, d^3x + \int_S u_i \sigma_{ij} n_j \, dS \\
&\quad + \int_S -\vec{K}\cdot\vec{n}\, dS + \int_R \rho Q \, d^3x
\end{aligned}
$$

($e = c_V T$, c_V = heat capacity at constant volume, Q = external heating, and $\vec{K}$ = heat flux). In view of the arbitrariness of R,

$$\rho \frac{D}{Dt}\left(\frac{u^2}{2} + e\right) = \rho f_i u_i + \frac{\partial}{\partial x_j}(\sigma_{ij} u_i) - \nabla \cdot \vec{K} + \rho Q.$$

Subtract from the above the relation for kinetic energy:

$$\rho \frac{De}{Dt} = \sigma_{ij}\frac{\partial u_i}{\partial x_j} - \nabla \cdot \vec{K} + \rho Q. \tag{6.8}$$

Equations 6.2, 6.4, 6.6, and 6.8 are our equations of motion – so far.

6.4 $\vec{K}$ and σ_{ij}

The nature of $\vec{K}$ and σ_{ij} is usually (and properly) discussed in terms of molecular collisions and/or turbulent mixing. We will take a somewhat different approach here. Let us assume

$$\vec{K} = \vec{K}\left(\rho, T, \frac{\partial T}{\partial x_i}\right).$$

Taylor expanding in $\frac{\partial T}{\partial x_i}$ we get

$$K_i = K_i^{(0)}(\rho, T) + A_{ij}(\rho, T)\frac{\partial T}{\partial x_j} + \cdots .$$

In general, $K_i^{(0)} = 0$. If we also assume transport to be *isotropic* then A_{ij} must be proportional to δ_{ij}; that is,

$$A_{ij} = -k(\rho, T)\delta_{ij},$$

and

$$\vec{K} = -k(\rho, T)\nabla T. \tag{6.9}$$

This result is, of course, far more convincingly obtained by kinetic theory considerations.

Following a similarly abstract approach for σ_{ij} we will assume

$$\sigma_{ij} = \sigma_{ij}\left(\rho, T, \frac{\partial u_k}{\partial x_l}\right).$$

(Fluids satisfying this assumption are known as *Newtonian* fluids.) Taylor expanding the above relation we get

$$\sigma_{ij} = B_{ij}(\rho, T) + C_{ijkl}(\rho, T)\frac{\partial u_k}{\partial x_l} + \cdots .$$

Again we want B_{ij} and C_{ijkl} to be isotropic tensors.

$$B_{ij} = - \underbrace{p}_{pressure}\, \delta_{ij}$$

$$C_{ijkl} = \lambda\delta_{ij}\delta_{kl} + \mu\underbrace{(\delta_{ik}\delta_{jl} + \delta_{jk}\delta_{il})}_{symmetric\ in\ i,j} + \nu\underbrace{(\delta_{ik}\delta_{jl} - \delta_{jk}\delta_{il})}_{antisymmetric\ in\ i,j}$$

Since $\sigma_{ij} = \sigma_{ji}$, $\nu \equiv 0$ and

$$\sigma_{ij} = -p\delta_{ij} + \lambda\delta_{ij}\nabla \cdot \vec{u} + \underbrace{\mu\left(\frac{\partial u_i}{\partial x_j} + \frac{\partial u_j}{\partial x_i}\right)}_{\tau_{ij},\ the\ viscous\ stress\ tensor}, \qquad (6.10)$$

where $\mu \equiv$ first viscosity and $\lambda \equiv$ second viscosity. The expression for τ_{ij} is a little more complicated than the most common expressions. Stokes suggested that average normal viscous stress should be zero; that is, $\tau_{ii} = 3\lambda\nabla \cdot \vec{u} + 2\mu\nabla \cdot \vec{u}$, which implies $\lambda + (2/3)\mu = 0$. The quantity $\eta = \lambda + (2/3)\mu$ is called the *bulk viscosity* and is zero for spherical molecules – but not otherwise. Still, for incompressible fluids where $\nabla \cdot \vec{u} \equiv 0$, the second viscosity is irrelevant. Also, if μ and λ are constant

$$\frac{\partial}{\partial x_j}\tau_{ij} = \lambda\frac{\partial}{\partial x_i}(\nabla \cdot \vec{u}) + \mu\left(\frac{\partial^2 u_i}{\partial x_j\,\partial x_j} + \frac{\partial^2 u_j}{\partial x_i\,\partial x_j}\right)$$

or

$$\hat{\imath}_i\frac{\partial \tau_{ij}}{\partial x_j} = (\lambda + \mu)\nabla(\nabla \cdot \vec{u}) + \underbrace{\mu\nabla^2\vec{u}}_{most\ commonly\ considered\ term}$$

With (6.10), $\sigma_{ij}\frac{\partial u_i}{\partial x_j}$ in (6.8) becomes

$$\sigma_{ij}\frac{\partial u_i}{\partial x_j} = -p\nabla \cdot \vec{u} + \Phi,$$

where $\Phi \equiv \tau_{ij}\frac{\partial u_i}{\partial x_j}$. Φ represents dissipation by viscosity. It can be shown that $\Phi \geq 0$ if and only if $\mu \geq 0$, $\eta \geq 0$. Using (6.9) and (6.10), and replacing $f_i \hat{i}_i$ with $\vec{g}$, (6.4) and (6.8) become

$$\rho\frac{Du_i}{Dt} = \rho g_i - \frac{\partial p}{\partial x_i} + \frac{\partial \tau_{ij}}{\partial x_j} \tag{6.11}$$

and

$$\rho c_V \frac{DT}{Dt} + p\nabla \cdot \vec{u} = \nabla \cdot (k\nabla T) + \rho Q + \Phi. \tag{6.12}$$

Note that Φ is generally small.

6.5   Equations of state

Equations 6.2, 6.11, and 6.12 represent five equations in six unknowns u_i, ρ, p, T. The remaining equation is the *equation of state*. Several choices are commonly made.

$$\text{Perfect gas:} \quad p = \rho RT \tag{6.13}$$
$$\text{Boussinesq fluid:} \quad \rho = \rho_0(1 - \alpha(T - T_0)) \tag{6.14}$$
$$\text{homog., incomp. fluid:} \quad \rho = \text{constant.} \tag{6.15}$$

In Equation 6.12 it is common to rewrite the left-hand side as follows:

$$\rho c_V \frac{DT}{Dt} + p\nabla \cdot \vec{u} = \rho c_V \frac{DT}{Dt} - \frac{p}{\rho}\frac{D\rho}{Dt} = \rho T\frac{DS}{Dt},$$

where S is entropy; note that thermodynamic equilibrium has been assumed. For a perfect gas

$$\frac{DS}{Dt} = \frac{c_v}{T}\frac{DT}{Dt} - \frac{R}{\rho}\frac{D\rho}{Dt}$$

$$= \frac{D}{Dt}\ln\left(\frac{T^{c_v}}{\rho^R}\right). \tag{6.16}$$

The relation between S and *potential temperature* ($\Theta \equiv T^{c_v/R}/\rho$) is obvious. For the Boussinesq fluid, S is proportional to T (or ρ), while for an incompressible, homogeneous fluid S is irrelevant.

6.6 Rotating coordinate frame

Consider a rotating frame (x', y', z'). A given position $\vec{r}$ can be represented

$$\vec{r} = x(t)\hat{i} + y(t)\hat{j} + z(t)\hat{k} \qquad \text{(inertial frame)}$$
$$= x'\hat{i}' + y'\hat{j}' + z'\hat{k}' \qquad \text{(rotating frame)}.$$

The velocity in the inertial frame can be written

$$\vec{u} = \frac{D\vec{r}}{Dt}$$

$$= \underbrace{\frac{Dx'}{Dt}\hat{i}' + \frac{Dy'}{Dt}\hat{j}' + \frac{Dz'}{Dt}\hat{k}'}_{\text{velocity in rotating frame}}$$

$$+ \underbrace{x'\frac{D\hat{i}'}{Dt} + y'\frac{D\hat{j}'}{Dt} + z'\frac{D\hat{k}'}{Dt}}_{\text{velocity of particle fixed in rotating frame: } \vec{\omega}\times\vec{r}}$$

So,

$$\vec{u} = \vec{u}' + \vec{\omega} \times \vec{r}$$

and

$$\frac{D}{Dt} = \left(\frac{D}{Dt}\right)_{rot} + \vec{\omega} \times \; .$$

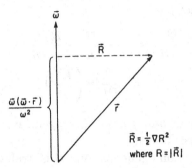

Figure 6.1: Centrifugal contribution to geopotential.

Thus for

$$\vec{u} = \vec{u'} + \vec{\omega} \times \vec{r},$$

$$
\begin{aligned}
\frac{D\vec{u}}{Dt} &= \left(\frac{D\vec{u}}{Dt}\right)_{rot} + \vec{\omega} \times \frac{D\vec{r}}{Dt} \\
&= \left(\frac{D}{Dt}\right)_{rot} (\vec{u'} + \vec{\omega} \times \vec{r}) + \omega \times (\vec{u'} + \vec{\omega} \times \vec{r}) \\
&= \left(\frac{D\vec{u'}}{Dt}\right)_{rot} + \vec{\omega} \times \vec{u'} + \vec{\omega} \times \vec{u'} + \vec{\omega} \times (\vec{\omega} \times \vec{r}).
\end{aligned}
$$

Now,

$$\vec{\omega} \times (\vec{\omega} \times \vec{r}) = (\vec{\omega} \cdot \vec{r})\vec{\omega} - \omega^2 \vec{r} = -\omega^2(\vec{r} - \frac{\vec{\omega}(\vec{\omega} \cdot \vec{r})}{\omega^2}) = -\omega^2 \vec{R}$$

$$(viz \text{ Figure 6.1}).$$

Thus, in a rotating frame, (6.11) becomes

$$\rho\left(\frac{D\vec{u'}}{Dt} + \overbrace{2\vec{\omega} \times \vec{u'}}^{Coriolis\ force}\right) = -\nabla p + \nabla \cdot \vec{\tau}$$

$$+ \underbrace{\rho\vec{g} + \rho\nabla\left(\frac{\omega^2 R^2}{2}\right)}_{usually\ combined\ as\ -\rho\nabla\Omega} . \quad (6.17)$$

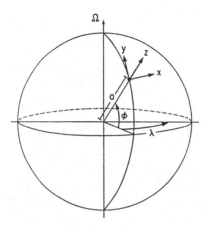

Figure 6.2: Spherical coordinates with cartesian tangent plane.

where $\Omega \equiv$ *geopotential*. Note that, because of isotropy, the viscous term is the same in rotating and non-rotating systems.

6.7 Spherical coordinates

We shall not derive the spherical equations here. The task is straightforward. The most direct approach involves transforming our equations into vector-invariant form (note that $\vec{u} \cdot \nabla$ is *not* vector-invariant), and then employing spherical forms for the invariant operations. Here we will merely write down equations discussed in Holton (his section 2.3; N.B. Holton uses an approach different from what we have just described). Holton considers a quasi-cartesian system on the surface of the earth (see Figure 6.2).

Sphere	Cartesian tangent plane	
$r = a + z$	dz	$w = \frac{dz}{dt}$
λ	$dx = a \cos\phi \, d\lambda$	$u = a \cos\phi \frac{d\lambda}{dt}$
ϕ	$dy = a \, d\phi$	$v = a \frac{d\phi}{dt}$

If we define

$$\frac{d}{dt} \equiv \frac{\partial}{\partial t} + u \frac{\partial}{\partial x} + v \frac{\partial}{\partial y} + w \frac{\partial}{\partial z},$$

then the equations for u,v, and w become

$$\frac{du}{dt} - \frac{uv\tan\phi}{a} + \frac{uw}{a} = -\frac{1}{\rho}\frac{\partial p}{\partial x} + 2\Omega v\sin\phi$$

$$-2\Omega w\cos\phi + \frac{1}{\rho}(\nabla\cdot\tau)_x \quad (6.18)$$

$$\frac{dv}{dt} + \frac{u^2\tan\phi}{a} + \frac{vw}{a} = -\frac{1}{\rho}\frac{\partial p}{\partial y} - 2\Omega u\sin\phi + \frac{1}{\rho}(\nabla\cdot\tau)_y \quad (6.19)$$

$$\frac{dw}{dt} + \frac{u^2+v^2}{a} = -\frac{1}{\rho}\frac{\partial p}{\partial z} - g + 2\Omega u\cos\phi + \frac{1}{\rho}(\nabla\cdot\tau)_z . \quad (6.20)$$

(What about equations for energy, continuity, and the gas law?)

6.8   Scaling

Scaling is an approach to non-dimensionalization which ideally permits ordering terms in an equation according to their size. For any variable, f, we may write

$$f = F\tilde{f},$$

where

$$
\begin{aligned}
F &= \text{characteristic magnitude} \\
\tilde{f} &= O(1)\ \text{non-dimensional quantity} .
\end{aligned}
$$

Such a redefinition leads to non-dimensional parameters which indicate the relative importance of various terms in equations. Some of the more famous non-dimensional numbers, and the physical balances they represent, are

$$
\begin{aligned}
\text{Rossby No.:}\quad & Ro = \frac{U}{2\Omega L} \sim \frac{\text{inertia}}{\text{Coriolis force}} \\[4pt]
\text{Froude No.:}\quad & Fr = \frac{gL}{U^2} \sim \frac{\text{gravity}}{\text{inertia}} \\[4pt]
\text{Reynolds No.:}\quad & Re = \frac{LU}{\nu} \sim \frac{\text{inertia}}{\text{friction}} \\[4pt]
\text{Ekman No.:}\quad & Ek = \frac{Ro}{Re} = \frac{\nu}{2\Omega L^2} \sim \frac{\text{friction}}{\text{Coriolis force}} ,
\end{aligned}
$$

(where $\nu \equiv \mu/\rho$). Note that there is little sense in scaling pressure in the absence of either data or the size of other numbers.

The following are some serious problems with scaling:

1. Northerly, westerly, and vertical scales and velocities are not necessarily the same.

2. Several scales for a single variable may be present in a single problem.

3. Not all variables are measured.

Ultimately, scaling actually requires that we already have knowledge of the answer! The following sections deal with some specific examples of scaling.

6.9 Hydrostaticity

As already noted, *hydrostaticity* is, in general, an approximation. We will use scaling to estimate the conditions needed for hydrostaticity to be a good approximation. One is tempted to note that g turns out to be much larger than any acceleration or stress in Equation 6.20 – at least for most meteorological and oceanographic systems. Thus, one might argue that the only term left to balance g is $\frac{1}{\rho}\frac{\partial p}{\partial z}$. This would, however, be a possibly misleading statement. To see why, let us write

$$
\begin{aligned}
p &= p_0(z) + p'(x, y, z, t) \\
\rho &= \rho_0(z) + \rho'(x, y, z, t) .
\end{aligned}
$$

The quantities p_0 and ρ_0 represent horizontal and time averages; the primed quantities represent deviations from these averages. In general, the averaged quantities dominate the above expressions; however, for stable stratification[3], the averaged terms are not directly involved in the dynamics (which depends on gradients of pressure, density, etc.). Scaling must be done *after* p_0 and ρ_0 (and T_0) have been subtracted from p and ρ (and T). The evaluation of the term $\rho' g$ in Equation 6.20

---

[3]We'll explain what this means soon.

then requires knowing how ρ' is related to w, and so forth. It is here that one has to involve the energy equation in the scaling, and one discovers that a necessary condition for hydrostaticity to hold is that $N\tau \gg 1$ (where N is the *Brunt-Vaisala* frequency and τ is a characteristic time scale).

This is most easily seen for a linearized adiabatic Boussinesq fluid where (6.20) becomes

$$\rho \frac{\partial w}{\partial t} \sim -g\rho - \frac{\partial p}{\partial z}. \tag{6.21}$$

Now, $\rho = \rho_0 + \rho'$, and $\frac{\partial p_0}{\partial z} = -g\rho_0$. Subtracting these from (6.21), we get for our vertical momentum equation

$$\rho_0 \frac{\partial w}{\partial t} \sim -g\rho' - \frac{\partial p'}{\partial z}.$$

Similarly, the Boussinesq energy equation becomes

$$\frac{\partial \rho'}{\partial t} + w \frac{d\rho_0}{dz} \sim 0.$$

For convenience, let our variables have the following time dependence: $e^{i\omega t}$. The last two equations become

$$\rho_0 i\omega w \sim -g\rho' - \frac{\partial p'}{\partial z}$$

$$i\omega \rho' \sim -w \frac{d\rho_0}{dz}.$$

Eliminating w from the above equations yields

$$\left(\frac{\rho_0 \omega^2}{\frac{d\rho_0}{dz}} \right) \rho' \sim -g\rho' - \frac{\partial p'}{\partial z},$$

from which we see that hydrostaticity requires that

$$\left| \frac{\rho_0 \omega^2}{g \frac{d\rho_0}{dz}} \right| \ll 1,$$

or

$$\left| \frac{\omega^2}{N^2} \right| \ll 1,$$

where $N^2 \equiv -\frac{g}{\rho_0} \frac{d\rho_0}{dz}$. The quantity N is called the Brunt-Vaisala frequency and is a measure of the stratification. When $N^2 < 0$, the fluid is statically unstable (that is, we have heavier fluid on top of lighter fluid)[4].

6.10   Geostrophy

In terms of scaling, one commonly finds that $Ro \ll 1$ and $Ek \ll 1$ so that pressure gradients primarily balance the Coriolis force. We also saw this directly in the data. Nevertheless, the existence of this approximate balance, alone, does not permit us to calculate the time evolution of flow fields. In order to exploit $Ro \ll 1$ in order to calculate the time evolution of almost geostrophic fields we will have to perform a more sophisticated scaling analysis (not too different from what we just did in connection with hydrostaticity) which leads to what are called the *quasi-geostrophic equations*. We will defer this analysis until a later chapter.

Before ending this chapter, an important aspect of geostrophic balance needs to be mentioned and pondered: namely, in the absence of viscosity, thermal conductivity, and x-variation, the following geostrophic flow is an exact solution of the nonlinear equations:

$T(y, z)$ as determined from $Q = 0$; that is, radiative equilibrium;

p, ρ as determined from hydrostaticity and the gas law;

u as determined from geostrophy; and

$v = w$ $= 0$.

The intrinsic difficulties with the above solution will form the focus for much of our discussion in Chapter 7.

---

[4]We will have much more to say about both the Brunt-Vaisala frequency and static stability in later chapters.

Chapter 7

Symmetric circulation models

Supplemental reading:

Lorenz (1967)

Held and Hou (1980)

Schneider and Lindzen (1977)

Schneider (1977)

Lindzen and Hou (1988)

As we noticed in our perusal of the data, atmospheric fields are far from being zonally symmetric. Some of the deviation from symmetry is forced by the inhomogeneity of the earth's surface, and some is autonomous (travelling cyclones, for example). Nevertheless, the zonally averaged circulation has, over the centuries, been the object of special attention. Indeed, the term 'general circulation' is frequently taken to mean the zonally averaged behavior. This is the viewpoint of Lorenz (1967).

   You are urged to read chapters 1, 3, and 4 of Lorenz. Chapter 1 is a short and especially insightful discussion of the methodology of studying the atmosphere. As is generally the case in this field, there

100

will be views in Lorenz which are not universally agreed on, but this hardly diminishes its value.

There are several reasons for focussing on the zonally averaged circulation:

1. Significant motion systems like the tropical tradewinds are well described by zonal averages.

2. The circulation of the atmosphere is only a small perturbation on a rigidly rotating basic state which is zonally symmetric.

3. The zonally averaged circulation is a convenient subset of the total circulation.

Our approach in this chapter will be to inquire how the atmosphere would behave in the absence of eddies. It is hoped that a comparison of such results with observations will lend some insight into what maintains the observed zonally averaged state. In particular, discrepancies may point to the rôle of eddies in maintaining the zonal average. This has been a matter of active controversy to the present.

7.1 Historical review

A very complete historical treatment of this subject is given in chapter 4 of Lorenz. We will only present a limited sketch here. The first treatment of contemporary relevance was that of Hadley (1735). Hadley's aim was to explain the easterly (actually northeasterly in the Northern Hemisphere) tradewinds of the tropics and the prevailing westerlies of middle latitudes. His brief explanation is summarized in Figure 7.1. Ignoring Hadley's error in assuming conservation of velocity rather than angular momentum, Hadley's argument ran roughly as follows:

1. Warm air rises at the equator and flows poleward at upper levels approximately conserving angular momentum. (In view of the remarks at the end of Chapter 6, it is not, however, at all obvious why one would have a meridional circulaton at all. We will discuss this later.) Because the distance from the axis of rotation diminishes with increasing latitude, large westerly currents are produced at high latitudes.

Figure 7.1: A schematic representation of the general circulation of the atmosphere as envisioned by Hadley (1735).

2. The westerly currents would be far larger than observed. It is, therefore, presumed that friction would reduce westerly currents. As a result, the return flow at the surface will have a momentum deficit leading to tropical easterlies.

The above model was generally accepted for over a century. The main criticism of this model was that it predicted northwesterly winds at midlatitudes whereas nineteenth century data suggested southwesterly winds (in the Northern Hemisphere). This difficulty was answered independently by Ferrel (1856) and Thomson (1857). Their hypothesized solution is shown in Figure 7.2. Briefly, Ferrel and Thomson supplemented Hadley's arguments as follows. They noted that at the latitude at which zonal flow is zero, there must be a maximum in pressure (Why?), and that within the frictional layer next to the surface, a shallow flow will be established down pressure gradients leading to the reversed cell shown in Figure 7.2 (today referred to as a Ferrel cell). *By allowing the Hadley circulation to remain at upper levels, the tropics can continue to supply midlatitudes with angular momentum* which presumably is communicated to the Ferrel cell by friction.

Although there was a general acceptance of the Ferrel-Thomson model of the general circulation in the late nineteenth century, there was also a general uneasiness due to the obvious fact that the observed circulation was not zonally symmetric. Moreover, all the models we have discussed were developed in only a qualitative, verbal way. This is, of course, not surprising since the quantitative knowledge of atmospheric heating, turbulent transfer, and so forth, was almost completely lacking. So was virtually any information about the atmosphere above the surface – except insofar as cloud motions indicated upper level winds.

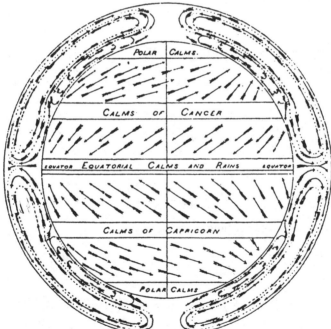

Figure 7.2: The general circulation of the atmosphere according to Thomson (1857).

In 1926, Jeffreys put forth an interesting and influential criticism of all symmetric models as an explanation of midlatitude surface westerlies. He began with an equation for zonal momentum (*viz.* Equation 6.18):

$$\frac{\partial u}{\partial t} + u\frac{\partial u}{\partial x} + v\frac{\partial u}{\partial y} + w\frac{\partial u}{\partial z} - \frac{uv\tan\phi}{a} + \frac{uw}{a}$$

$$= -\frac{1}{\rho}\frac{\partial p}{\partial x} + 2\Omega v\sin\phi - 2\Omega w\cos\phi + \frac{1}{\rho}(\nabla\cdot\tau)_x. \qquad (7.1)$$

For the purposes of Jeffreys' argument, (7.1) can be substantially simplified. Steadiness and zonal symmetry (no eddies) imply $\frac{\partial}{\partial t} = \frac{\partial}{\partial x} = 0$. Scaling shows that

$$\frac{uw}{a}, 2\Omega w\cos\phi \ll \frac{uv\tan\phi}{a}, 2\Omega v\sin\phi$$

(What about the equator?). In addition,

$$v\frac{\partial u}{\partial y} - \frac{uv\tan\phi}{a} = \frac{v}{a}\frac{\partial u}{\partial\phi} - \frac{uv\tan\phi}{a} = \frac{v}{a\cos\phi}\frac{\partial}{\partial\phi}(u\cos\phi).$$

Thus (7.1) becomes

$$\frac{v}{a\cos\phi}\frac{\partial}{\partial\phi}(u\cos\phi) + w\frac{\partial u}{\partial z} - 2\Omega v\sin\phi = \frac{1}{\rho}(\nabla\cdot\tau)_x. \qquad (7.2)$$

Jeffreys further set

$$(\nabla\cdot\tau)_x = \frac{\partial}{\partial z}\mu\frac{\partial u}{\partial z}. \qquad (7.3)$$

Integrating (7.2) over all heights yields

$$\int_0^\infty \frac{\rho v}{a\cos\phi}\frac{\partial}{\partial\phi}(u\cos\phi)\,dz + \int_0^\infty \rho w\frac{\partial u}{\partial z}\,dz$$

$$-2\Omega\sin\phi\int_0^\infty \rho_0 v\,dz = -\mu\left.\frac{\partial u}{\partial z}\right|_0 = -C_D u_0^2, \qquad (7.4)$$

where C_D is the surface drag coefficient, and $C_D u_0^2$ is the usual phenomenological expression for the surface drag. Note that the earth's angular momentum cannot supply momentum removed by surface drag (since there is no net meridional mass flow; i.e., $\int_0^\infty \rho_0 v \, dz = 0.$). Thus, $C_D u_0^2$ must be balanced by the advection of relative momentum. Jeffreys argued that the integrals on the left-hand side of (7.4) would be dominated be the first integral evaluated within the first kilometer or so of the atmosphere. Underlying his argument was the obvious lack of data to do otherwise. He then showed that this integral was about a factor of 20 smaller than $C_D u_0^2$. He concluded that the maintenance of the surface westerlies had to be achieved by the neglected eddies. It may seem odd that Jeffreys, who so carefully considered the effect of the return v-flow on the Coriolis torque, ignored it for the transport of relative momentum. However, since it was the Ferrel cell he was thinking of, its inclusion would not have altered his conclusion. What he failed to note was that in both Hadley's model and that of Ferrel and Thomson, it was the Hadley cell which supplied westerly momentum to middle latitudes. Thus Jeffreys' argument is totally inconclusive; it certainly is not a proof that a symmetric circulation would be impossible (though this was sometimes claimed in the literature).

A more balanced view was presented by Villem Bjerknes (1937) towards the end of his career. Bjerknes suggested that in the absence of eddies the atmosphere would have a Ferrel-Thomson circulation – but that such an atmosphere would prove unstable to eddies. This suggestion did not, however, offer any estimate of the extent to which the symmetric circulation could explain the general circulation, and the extent to which eddies are essential.

In recent years Ed Schneider and I have attempted to answer this question by means of a rather cumbersome numerical calculation (Schneider and Lindzen, 1977; Schneider, 1977). The results shown in Figure 7.3 largely confirm Bjerknes' suggestion. The main shortcoming of this calculation was that it yielded a zonal jet that was much too strong. Surface winds were also a little weaker than observed, but on the whole, the symmetric circulation suffered from none of the inabilities Jeffreys had attributed to it. In order to see how the symmetric circulation works, it is fortunate that Schneider (1977) discovered a rather simple approximate approach to calculating the Hadley

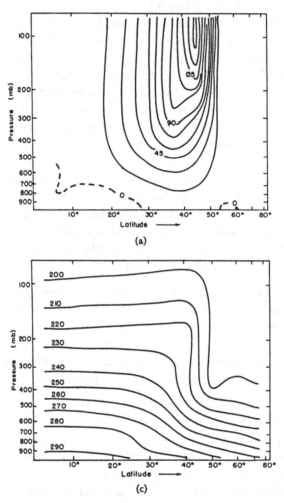

Figure 7.3: Example of an eddy-free symmetric circulation found by Schneider (1977). Panel (a) shows contours of zonal wind (contour intervals of $15\,\mathrm{ms}^{-1}$). Panel (b) shows streamfunction contours (contour intervals of $10^{12}\ \mathrm{gs}^{-1}$). Panel (c) shows temperature contours (contour intervals of $10\,\mathrm{K}$).

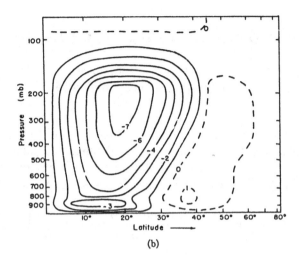

(b)

circulation. Held and Hou (1980) explored this approximation in some detail. We shall briefly go over the Held and Hou calculations.

7.2   Held and Hou calculations

Held and Hou restrict themselves to a Boussinesq fluid of depth H. For such a fluid, the continuity equation is simplified to

$$\nabla \cdot \vec{u} = 0. \tag{7.5}$$

With (7.5) as well as the assumptions of steadiness and zonal symmetry, and the retention of only vertical diffusion in the viscous stress and thermal conduction terms, our remaining equations of motion become

$$\nabla \cdot (\vec{u}u) - \underbrace{f}_{2\Omega \sin \phi} v - \frac{uv \tan \phi}{a} = \frac{\partial}{\partial z}\left(\nu \frac{\partial u}{\partial z}\right) \tag{7.6}$$

$$\nabla \cdot (\vec{u}v) + fu + \frac{u^2 \tan \phi}{a} = -\frac{1}{a}\frac{\partial \Phi}{\partial \phi} + \frac{\partial}{\partial z}\left(\nu \frac{\partial v}{\partial z}\right) \tag{7.7}$$

$$\nabla \cdot (\vec{u}\Theta) = \frac{\partial}{\partial z}\left(\nu \frac{\partial \Theta}{\partial z}\right) - \frac{(\Theta - \Theta_E)}{\tau} \tag{7.8}$$

and

$$\frac{\partial \Phi}{\partial z} = g\frac{\Theta}{\Theta_0}. \tag{7.9}$$

(N.B. $\Phi = \frac{p}{\rho}$.) The quantity Θ_E is presumed to be a 'radiative' equilibrium temperature distribution for which we adopt the simplified form

$$\frac{\Theta_E(\phi, z)}{\Theta_0} \equiv 1 - \Delta_H\left(\frac{1}{3} + \frac{2}{3}P_2(\sin \phi)\right) + \Delta_V\left(\frac{z}{H} - \frac{1}{2}\right), \tag{7.10}$$

where $P_2(x) \equiv \frac{1}{2}(3x^2 - 1)$, $x = \sin \phi$, $\Theta_0 = \Theta_E(0, \frac{H}{2})$, Δ_H = fractional potential temperature drop from the equator to the pole, Δ_V = fractional potential temperature drop from H to the ground, and τ is a 'radiative' relaxation time. (The reader should work out the derivation of Equations 7.6–7.9. N.B. In the Boussinesq approximation, density is taken as constant except where it is multiplied by g. The resulting simplifications can also be obtained for a fully stratified atmosphere by using the log-pressure coordinates described in Chapter 4.)

The boundary conditions employed by Held and Hou are

$$\underbrace{\frac{\partial u}{\partial z} = \frac{\partial v}{\partial z}}_{no\ stress} = \overbrace{\frac{\partial \Theta}{\partial z}}^{no\ heat\ conduction} = \underbrace{w}_{rigid\ top} = 0 \qquad \text{at } z = H \qquad (7.11)$$

$$\frac{\partial \Theta}{\partial z} = w = 0 \qquad \text{at } z = 0 \qquad (7.12)$$

$$\underbrace{\nu \frac{\partial u}{\partial z} = Cu, \quad \nu \frac{\partial v}{\partial z} = Cv}_{linearization\ of\ surface\ stress\ conditions} \qquad \text{at } z = 0 \qquad (7.13)$$

$$\underbrace{v = 0 \qquad \text{at } \phi = 0}_{symmetry\ about\ the\ equator} . \qquad (7.14)$$

The quantity C is taken to be a constant drag coefficient. Note that this is not the same drag coefficient that appeared in (7.4); neither is the expression for surface drag which appears in (7.13) the same. As noted in (7.13), the expression is a linearization of the full expression. The idea is that the full expression is quadratic in the *total* surface velocity – of which the contribution of the Hadley circulation is only a part. The coefficient C results from the product of C_D and the 'ambient' surface wind.

When $\nu \equiv 0$ we have already noted that our equations have an exact solution:

$$v = w = 0 \qquad (7.15)$$
$$\Theta = \Theta_E \qquad (7.16)$$

and

$$u = u_E, \qquad (7.17)$$

where u_E satisfies

$$\frac{\partial}{\partial z}\left(fu_E + \frac{u_E^2 \tan\phi}{a}\right) = -\frac{g}{a\Theta_0}\frac{\partial\Theta_E}{\partial\phi} \qquad \text{(where } y = a\phi\text{).} \qquad (7.18)$$

If we set $u_E = 0$ at $z = 0$, the appropriate integral of (7.18) is (see Holton, 1979, p. 62):

$$\frac{u_E}{\Omega a} = \left[\left(1 + 2R\frac{z}{H}\right)^{1/2} - 1\right]\cos\phi, \qquad (7.19)$$

where

$$R = \frac{gH\Delta_H}{(\Omega a)^2}. \qquad (7.20)$$

When $R \ll 1$,

$$\frac{u_E}{\Omega a} = R\cos\phi\frac{z}{H}. \qquad (7.21)$$

Why isn't the above solution, at least, approximately appropriate? Why do we need a meridional solution at all? Hadley already implicitly recognized that the answer lies in the presence of viscosity. A theorem (referred to as 'Hide's theorem') shows that if we have viscosity (no matter how small), (7.15)–(7.19) cannot be a steady solution of the *symmetric* equations.

7.2.1   Hide's theorem and its application

The proof of the theorem is quite simple. We can write the total angular momentum per unit mass as

$$M \equiv \Omega a^2 \cos^2\phi + ua\cos\phi \qquad (7.22)$$

(recall that ρ is taken to be 'constant'), and (7.6) may be rewritten

$$\nabla \cdot (\vec{u}M) = \nu\frac{\partial^2 M}{\partial z^2}. \qquad (7.23)$$

Now suppose that M has a local maximum somewhere in the fluid. We may then find a closed contour surrounding this point where M is constant. If we integrate (7.23) about this contour, the contribution of the left-hand side will go to zero (Why?), while the contribution of the right-hand side will be negative (due to down gradient viscous fluxes). Since such a situation is inconsistent, M cannot have a maximum in the interior of the fluid. We next consider the possibility that M has a maximum at the surface. We may now draw a constant M contour above the surface, and close the contour along the surface (where $w = u_n = 0$). Again, the contribution from the left-hand side will be zero. The contribution from the right-hand side will depend on the sign of the surface wind. If the surface wind is westerly, then the contribution of the right-hand side will again be negative, and M, therefore, cannot have a maximum at the surface where there are surface westerlies. If the surface winds are easterly, then there is, indeed, a possibility that the contribution from the right-hand side will be zero. Thus, the maximum value of M must occur at the surface in a region of surface easterlies[1]! An upper bound for M is given by its value at the equator when $u = 0$; that is,

$$M_{max} < \Omega a^2. \tag{7.24}$$

Now u_E as given by (7.19) implies (among other things) westerlies at the equator and increasing M with height at the equator – all of which is forbidden by *Hide's theorem* – at least for symmetric circulations. *A meridional circulation is needed in order to produce adherence to Hide's theorem.*

Before proceeding to a description of this circulation we should recall that in our discussion of observations we did indeed find zonally averaged westerlies above the equator (in connection with the quasi-biennial oscillation, for example). This implies the existence of eddies which are transporting angular momentum up the gradient of mean angular momentum!

---

[1]It is left to the reader to show that M cannot have a maximum at a stress-free upper surface.

A clue to how much of a Hadley circulation is needed can be obtained by seeing where $u_E = u_M$; by u_M we mean the value of u associated with $M = \Omega a^2$ (*viz* Equation 7.24). From (7.22) we get

$$u_M = \frac{\Omega a \sin^2 \phi}{\cos \phi}. \tag{7.25}$$

Setting $u_E = u_M$ gives an equation for $\phi = \phi^*$. For $\phi < \phi^*$, u_E violates Hide's theorem.

$$[(1 + 2R)^{1/2} - 1] \cos \phi^* = \frac{\sin^2 \phi^*}{\cos \phi^*}. \tag{7.26}$$

(Recall that $R \equiv \frac{gH\Delta_H}{(\Omega a)^2}$.)

Solving (7.26) we get

$$\phi^* = \tan^{-1}\{[[(1 + 2R)^{1/2} - 1)]^{1/2}\}. \tag{7.27}$$

For small R,

$$\phi^* = R^{1/2}. \tag{7.28}$$

Using reasonable atmospheric values

$$g = 9.8 \text{ ms}^{-2}$$
$$H = 1.5 \ 10^4 \text{ m}$$
$$\Omega = 2\pi/(8.64 \ 10^4 \text{ s})$$
$$a = 6.4 \ 10^6 \text{ m}$$

and

$$\Delta_H \sim 1/3,$$

we obtain from (7.20)

$$R \approx .226$$

and

$$R^{1/2} \approx .48$$

and from (7.28)

$$\phi^* \approx 30°.$$

Thus, we expect a Hadley cell over at least half the globe[2].

7.2.2 Simplified calculations

Solving for the Hadley circulation is not simple even for the highly simplified model of Held and Hou. However, Schneider and Held and Hou discovered that the solutions they ended up with when viscosity was low were approximately constrained by a few principles which served to determine the main features of the Hadley circulation:

1. The upper poleward branch conserves angular momentum;

2. The zonal flow is balanced; and

3. Surface winds are small compared to upper level winds.
   In addition:

4. Thermal diffusion is not of dominant importance in Equation 7.8.

Held and Hou examine, in detail, the degree to which these principles are valid, and you are urged to read their work. However, here we shall merely examine the implications of items (1)–(4) and see how these compare with the numerical solutions from Held and Hou. Principle (1) implies

$$u(H, \phi) = u_M = \frac{\Omega a \sin^2 \phi}{\cos \phi}. \qquad (7.29)$$

---

[2]The choice $\Delta_H = 1/3$ is taken from Held and Hou (1980) and corresponds to Θ_E varying by about 100° between the equator and the poles. This, indeed, is reasonable for radiative equilibrium. However, more realistically, the atmosphere is, at any moment, more nearly in equilibrium with the sea surface (because adjustment times for the sea are much longer than for the atmosphere) and, therefore, a choice of $\Delta_H \approx 1/6$ may be more appropriate. This leads to $R^{1/2} \approx .34$ and $\phi^* \approx 20°$, which is not too different from what was obtained for $\Delta_H = 1/3$. This relative insensitivity of the Hadley cell extent makes it a fairly poor variable for distinguishing between various parameter choices.

Principle (2) implies

$$fu + \frac{u^2 \tan \phi}{a} = -\frac{1}{a}\frac{\partial \Phi}{\partial \phi}. \tag{7.30}$$

Evaluating (7.30) at $z = H$ and $z = 0$, and subtracting the results yields

$$f[u(H) - u(0)] + \frac{\tan \phi}{a}[u^2(H) - u^2(0)]$$
$$= -\frac{1}{a}\frac{\partial}{\partial \phi}[\Phi(H) - \Phi(0)]. \tag{7.31}$$

Integrating (7.9) from $z = 0$ to $z = H$ yields

$$\frac{\Phi(H) - \Phi(0)}{H} = \frac{g}{\Theta_0}\bar{\Theta}, \tag{7.32}$$

where $\bar{\Theta}$ is the vertically averaged potential temperature. Substituting (7.32) into (7.31) yields a simplified 'thermal wind' relation

$$f[u(H) - u(0)] + \frac{\tan \phi}{a}[u^2(H) - u^2(0)] = -\frac{gH}{a\Theta_0}\frac{\partial \bar{\Theta}}{\partial \phi}. \tag{7.33}$$

Principle (3) allows us to set $u(0) = 0$. Using this and (7.29), (7.33) becomes

$$2\Omega \sin \phi \frac{\Omega a \sin^2 \phi}{\cos \phi} + \frac{\tan \phi}{a}\frac{\Omega^2 a^2 \sin^4 \phi}{\cos^2 \phi} = -\frac{gH}{a\Theta_0}\frac{\partial \bar{\Theta}}{\partial \phi}. \tag{7.34}$$

Equation 7.34 can be integrated with respect to ϕ to obtain

$$\frac{\bar{\Theta}(0) - \bar{\Theta}(\phi)}{\Theta_0} = \frac{\Omega^2 a^2}{gH}\frac{\sin^4 \phi}{2\cos^2 \phi}. \tag{7.35}$$

Note that conservation of angular momentum and the maintenance of a balanced zonal wind completely determine the variation of $\bar{\Theta}$ within the Hadley regime. Moreover, the decrease of $\bar{\Theta}$ with latitude is much slower near the equator than would be implied by Θ_E!

Finally, we can determine both $\bar{\Theta}(0)$ and the extent of the Hadley cell, ϕ_H, with the following considerations:

1. At ϕ_H, temperature should be continuous so

$$\bar{\Theta}(\phi_H) = \bar{\Theta}_E(\phi_H). \tag{7.36}$$

2. From Equation 7.8 we see that the Hadley circulation does not produce net heating over the extent of the cell. For the diabatic heating law in Equation 7.8 we therefore have

$$\int_0^{\phi_H} \bar{\Theta} \cos\phi\, d\phi = \int_0^{\phi_H} \bar{\Theta}_E \cos\phi\, d\phi. \tag{7.37}$$

Substituting (7.35) into (7.36) and (7.37) yields the two equations we need in order to solve for ϕ_H and $\bar{\Theta}(0)$. The solution is equivalent to matching (7.35) to $\bar{\Theta}_E$ so that 'equal areas' of heating and cooling are produced. This is schematically illustrated in Figure 7.4. Also shown are $u_M(\phi)$ and $u_E(\phi)$.

The algebra is greatly simplified by assuming small ϕ. Then (7.35) becomes

$$\frac{\bar{\Theta}}{\Theta_0} \approx \frac{\bar{\Theta}(0)}{\Theta_0} - \frac{1}{2}\frac{\Omega^2 a^2}{gH}\phi^4 \tag{7.38}$$

and (7.10) becomes

$$\frac{\bar{\Theta}_E}{\Theta_0} = \frac{\bar{\Theta}_E(0)}{\Theta_0} - \Delta_H \phi^2. \tag{7.39}$$

Substituting (7.38) and (7.39) into (7.36) and (7.37) yields

$$\frac{\bar{\Theta}(0)}{\Theta_0} = \frac{\bar{\Theta}_E(0)}{\Theta_0} - \frac{5}{18} R\Delta_H \tag{7.40}$$

and

$$\phi_H = \left(\frac{5}{3}R\right)^{1/2}. \tag{7.41}$$

(Remember that $R \equiv \frac{gH\Delta_H}{\Omega^2 a^2}$; for a slowly rotating planet such as Venus, ϕ_H can extend to the pole.)

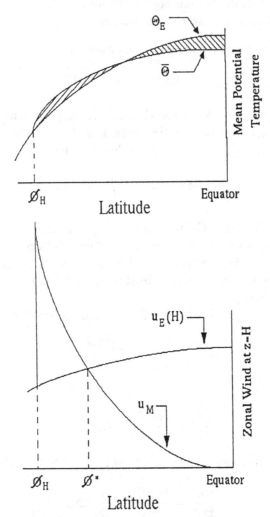

Figure 7.4: Schematic drawings of the vertical mean potential temperature distribution (upper figure) and the zonal wind distribution
at the top of the atmosphere (lower figure). With Newtonian cooling
(linear in Θ), conservation of potential temperature requires that the
shaded areas be equal. Note that this idealized circulation increases
the baroclinicity of the flow between ϕ^* (where $u_E = U_M$) and ϕ_H.

We see that continuity of potential temperature and conservation of angular momentum and potential temperature serve to determine the meridional distribution of temperature. The intensity of the Hadley circulation will be such as to produce this temperature distribution. In section 10.2 of Houghton (1977) there is a description of Charney's viscosity dominated model for a meridional circulation. In that model, $\Theta = \Theta_E$ and $u = u_E$, except in thin boundary layers, and the meridional velocity is determined by requiring that fv balance the viscous diffusion of momentum, $\nu \frac{\partial^2 u}{\partial z^2}$. Such a model clearly violates Hide's theorem. A more realistic viscous model is described by Schneider and Lindzen (1977) and Held and Hou (1980), wherein the meridional circulation is allowed to modify Θ through the following linearization of the thermodynamic energy equation

$$-w\frac{\partial \Theta}{\partial z} = (\Theta - \Theta_E)/\tau,$$

where $\frac{\partial \Theta}{\partial z}$ is a specified constant. In such models the meridional circulation continues, with gradual diminution, to high latitudes rather than ending abruptly at some subtropical latitude – as happens in the present 'almost inviscid' model. On the other hand, the modification of Θ (for the linear viscous model) is restricted to a neighbourhood of the equator given by

$$\phi \sim R^{*1/4},$$

where

$$R^* \equiv \left(\frac{\tau\nu}{4H}\right)\left(\frac{gH}{\Omega^2 a^2}\right)\Delta_V.$$

When $R^{*1/4} \geq R^{1/2}$ (*viz.* Equation 7.28) then a viscous solution can be compatible with Hide's theorem. Note, however, that linear models cannot have surface winds (Why? Hint: consider the discussion of Jeffreys' argument.).

Returning to our present model, there is still more information which can be extracted. Obtaining the vertically integrated flux of potential temperature is straightforward:

$$\frac{1}{H}\int_0^H \frac{1}{a\cos\phi}\frac{\partial}{\partial\phi}(v\Theta\cos\phi)\,dz = \frac{\bar{\Theta}_E - \bar{\Theta}}{\tau}. \tag{7.42}$$

In the small ϕ limit, $\bar{\Theta}_E$ and $\bar{\Theta}$ are given by (7.38)–(7.41) and (7.42) can be integrated to give

$$\frac{1}{\Theta_0}\int_0^H v\Theta\,dz$$

$$= \frac{5}{18}\left(\frac{5}{3}\right)^{1/2}\frac{Ha\Delta_H}{\tau}R^{3/2}\left[\frac{\phi}{\phi_H} - 2\left(\frac{\phi}{\phi_H}\right)^3 + \left(\frac{\phi}{\phi_H}\right)^5\right]. \tag{7.43}$$

Held and Hou are also able to estimate surface winds on the basis of this simple model. For this purpose additional assumptions are needed:

1. One must assume either
   (a) the meridional flow is primarily confined to thin boundary layers adjacent to the two horizontal boundaries, or that
   (b) profiles of u and Θ are self-similar so that

$$\frac{u(z) - u(0)}{u(H) - u(0)} \approx \frac{\Theta(z) - \Theta(0)}{\Theta(H) - \Theta(0)}.$$

   (We shall employ (a) because it's simpler.)

2. Neither the meridional circulation nor diffusion affects the static stability so that

$$\frac{\Theta(H) - \Theta(0)}{\Theta_0} \approx \Delta_V.$$

   (This requires that the circulation time and the diffusion time both be longer than τ; a serious discussion of this would require consideration of cumulus convection.)

With assumptions (1) and (2) above we can write

$$\frac{1}{\Theta_0}\int_0^H v\Theta\,dz \approx V\Delta_V, \tag{7.44}$$

where V is a mass flux in the boundary layers. With (7.44), (7.43) allows us to solve for V.

Similarly, we have for the momentum flux

$$\int_0^H vu\,dz \approx Vu_M. \tag{7.45}$$

To obtain the surface wind, we vertically integrate (7.23) (using (7.22) and (7.13)) to get

$$\frac{1}{a\cos^2\phi}\frac{\partial}{\partial\phi}\left(\cos^2\phi\int_0^H uv\,dz\right) = -Cu(0). \tag{7.46}$$

From (7.43)–(7.46) we then get

$$Cu(0) \approx -\frac{25}{18}\frac{\Omega a H \Delta_H}{\tau\Delta_V}R^2\left[\left(\frac{\phi}{\phi_H}\right)^2 - \frac{10}{3}\left(\frac{\phi}{\phi_H}\right)^4 + \frac{7}{3}\left(\frac{\phi}{\phi_H}\right)^6\right]. \tag{7.47}$$

Equation 7.47 predicts surface easterlies for

$$\phi < \left(\frac{3}{7}\right)^{1/2}\phi_H \tag{7.48}$$

and westerlies for

$$\left(\frac{3}{7}\right)^{1/2}\phi_H < \phi < \phi_H. \tag{7.49}$$

For the parameters given following Equation 7.28, the positions of the upper level jet and the easterlies and westerlies are moderately close to those observed. It can also be shown that for small ν, the above scheme leads to a Ferrel cell above the surface westerlies. We will return to this later. For the moment we wish to compare the results of the present simple analysis with the results of numerical integrations of Equations 7.5–7.14.

7.2.3   Comparison of simple and numerical results

Unfortunately, the results in Held and Hou are for $H = 8 \times 10^3$m rather than 1.5×10^4m. From our simple relations, we correctly expect this to

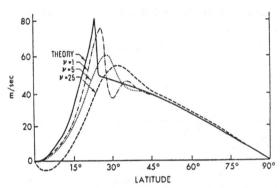

Figure 7.5: Zonal wind at $z = H$ for three values of ν, compared with the simple model for the limit $\nu \to 0$ (from Held and Hou, 1980).

cause features to be compressed towards the equator. Also, Held and Hou adopted the following values for τ, C, and Δ_V:

$$\tau = 20 \text{ days}$$
$$C = 0.005 \text{ ms}^{-1}$$
$$\Delta_V = 1/8.$$

Figures 7.5–7.8 compare zonal winds at $z = H$, M at $z = H$, heat flux, and surface winds for our simple calculations and for numerical integrations with various choices of ν. In general, we should note the following:

1. As we decrease ν the numerical results more or less approach the simple results, and for $\nu = .5 \text{ m}^2\text{s}^{-1}$ (generally accepted as a 'small' value) our simple results are a decent approximation. (In fact, however, reducing ν much more does not convincingly show that the limit actually is reached since the numerical solutions become unsteady.)

2. The presence of modest vertical viscosity increases and broadens both heat flux and the distribution of surface winds (Why?). Viscosity also reduces the magnitude of, broadens, and moves poleward the upper level jet.

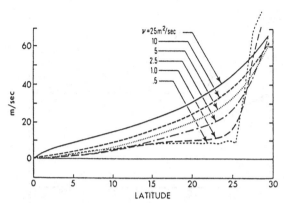

Figure 7.6: A measure of M, namely, $((\Omega a^2 - M)/a)$, evaluated at $z = H$ as a function of ϕ for diminishing values of viscosity, ν. Note that zero corresponds to conservation of M (from Held and Hou, 1980).

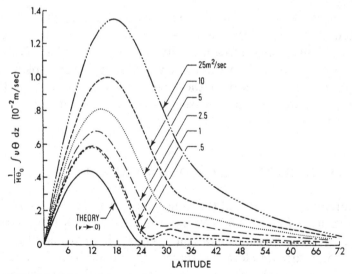

Figure 7.7: Meridional heat fluxes for various values of ν – as well as the theoretical limit based on the simple calculations (from Held and Hou, 1980).

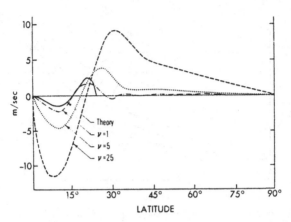

Figure 7.8: Surface wind for various values of ν, and the theoretical inviscid limit based on simple calculations (from Held and Hou, 1980).

3. Very near the equator, the numerical results do not quite converge to constant M at $z = H$. The reason for this can be seen in Figures 7.9 and 7.10 where meridional cross sections are shown for the meridional stream functions and zonal wind. Note that the upward branch of the Hadley cell does not rise solely at the equator (as supposed in the simple theory) but over a 10–15° neighbourhood of the equator. Note also the emergence of the Ferrel cell at small ν.

7.3   Summary and difficulties

Before summarizing what all this tells us about the general circulation let us return to Figure 7.3. We see that Schneider's symmetric circulation, which is by and large consistent with Held and Hou's, also manages to predict an elevated tropical tropopause height and the associated tropopause 'break' at the edge of the Hadley circulation. The midlatitude tropopause somewhat artificially reflects the assumed Θ_E distribution. The elevated tropopause in the tropics results from the inclusion of cumulus heating.

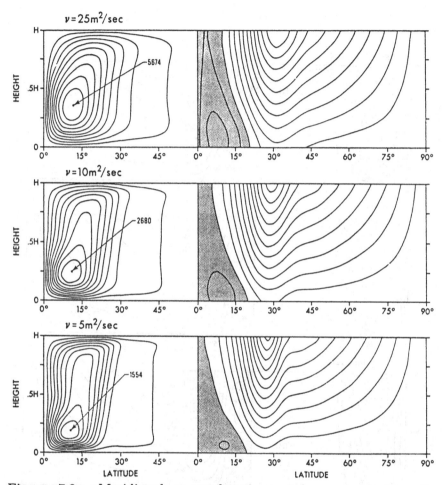

Figure 7.9: Meridional streamfunctions and zonal winds. In the left part of the figure, the streamfunction ψ is given for $\nu = 25, 10,$ *and* $5\,\mathrm{m^2s^{-1}}$, with a contour interval of 0.1 ψ_{max}. The value of 0.1 ψ_{max} $(\mathrm{m^2s^{-1}})$ is marked by a pointer. The right part of each panel is the corresponding zonal wind field with contour intervals of $5\,\mathrm{ms^{-1}}$. The shaded area indicates the region of easterlies (from Held and Hou, 1980).

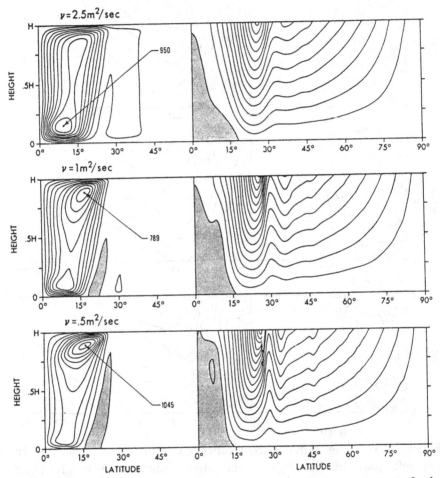

Figure 7.10: Same as Figure 7.9, but for $\nu = 2.5$, 1.0, *and* $0.5\,\mathrm{m^2s^{-1}}$. Note the emergence of a Ferrel cell in the ψ-field where $\psi < 0$ (indicated by shading) (from Held and Hou, 1980).

7.3.1 Remarks on cumulus convection

Cumulus heating will not be dealt with in these lectures, but for the moment three properties of cumulus convection should be noted:

1. It is, in practice, the primary mechanism for carrying heat from the surface in the tropics.

2. Cumulus towers, for simple thermodynamic reasons, extend as high as 16 km, and appear to be the determinant of the tropical tropopause height and the level of Hadley outflow. Remember that tropical circulations tend to wipe out horizontal gradients. Thus, the tropopause tends to be associated with the height of the deepest clouds.

3. Cumulus convection actively maintains a dry static stability (as required in the calculation of Hadley transport). This is explained in Sarachik (1985).

A more detailed description of cumulus convection (and its parameterization) can be found in Lindzen (1988b).

7.3.2 Tentative summary

On the basis of our study of symmetric circulations (so far) we find the following:

1. Symmetric solutions yield an upper level jet in about the right place but with much too *large* a magnitude.

2. Symmetric circulations yield surface winds of the right sign in about the right place. In the absence of vertical diffusion, magnitudes are too small, but modest amounts of vertical diffusion corrects matters, and cumulus clouds might provide this 'diffusion'. (Schneider and Lindzen, 1976, discuss cumulus friction.)

3. Calculated Hadley circulations have only a finite extent. In contrast to Hadley's and Ferrel and Thomson's models, the upper branch does not extend to the poles. Thus our Hadley circulation cannot carry heat between the tropics and the poles, and cannot produce the observed pole–equator temperature difference.

4. The calculated temperature distribution does not have the pronounced equatorial minimum at tropopause levels that is observed (*viz.* Figure 5.11).

5. Although not remarked upon in detail, the intensity of the Hadley circulation shown in Figure 7.10 is weaker than what is observed.

At this point we could glibly undertake to search for the resolution of the above discrepancies in the rôle of the thus far neglected eddies. However, before doing this, it is important to ask whether our symmetric models have not perhaps been inadequate in some other way besides the neglect of eddies.

7.4   Asymmetry about the equator

Although we do not have the time to pursue this (and most other matters) adequately, the reader should be aware that critical reassessments are essential to the scientific enterprise – and frequently the source of truly important problems and results. What is wrong with our results is commonly more important than what is right! A particular shortcoming will be discussed here: namely, the assumption that annual average results can be explained with a model that is symmetric about the equator. The importance of this shortcoming has only recently been recognized. (This section is largely based on the material in a paper by Lindzen and Hou, 1988.) This fact alone should encourage the reader to adopt a more careful and critical attitude.

What is at issue in the symmetry assumption can most easily be seen by looking at some data for the meridional circulation itself. Thus far we have not paid too much attention to this field. Figure 7.11 shows the meridional circulation for solstitial conditions. Not surprisingly, it is not symmetric about the equator. More surprising, however, is the degree of asymmetry: the 'winter' cell extends from well into the summer hemisphere ($\sim 20°$) to well into the winter hemisphere ($\sim 30°$), whereas the 'summer' cell barely exists at all! Figures 5.10 and 5.11 show meridional sections of zonally averaged zonal wind and temperature for solstitial conditions. Within about 20–25° of the equator these fields are symmetric about the equator. Thus, in this region, at least,

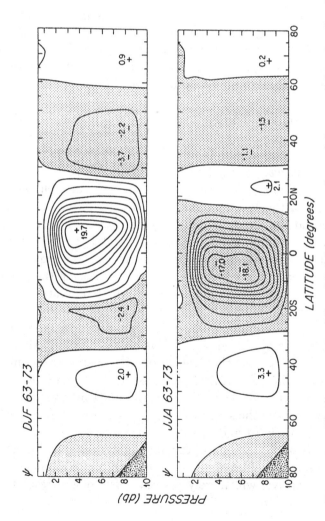

Figure 7.11: Time average meridional-height cross sections of the streamfunction for the mean meridional circulation. Units, 10^{13} gs^{-1}; contour intervals, 0.2×10^{13} gs^{-1}. December–February 1963–73 (upper panel) and June–August 1963–73 (lower panel) (from Oort, 1983).

an average over winter and summer of these fields will still give the solstitial distributions. Finally, Figure 7.12 shows the monthly means of the meridional circulation for each of the twelve months of the year. This figure is a little hard to interpret since it extends to only 15°S. However, to a significant extent it suggests that except for the month of April, the asymmetric solstitial pattern is more nearly characteristic of every month than is the idealized symmetric pattern invoked since Hadley in the eighteenth century. Clearly, the assumption of such symmetry is suspect.

Some insight into what is going on can, interestingly enough, be gotten from the simple 'equal area' argument. This is shown in Lindzen and Hou (1988). We will briefly sketch their results here. They studied the axially symmetric response to heating centered off the equator at some latitude ϕ_0. Thus Equation 7.10 was replaced by

$$\frac{\Theta_E}{\Theta_0} \cong 1 - \Delta_H \left(\sin\phi - \sin\phi_0\right)^2 + \Delta_V \left(\frac{z}{H} - \frac{1}{2}\right). \qquad (7.50)$$

The fact that $\phi_0 \neq 0$ substantially complicates the problem. Now the northward and southward extending cells will be different. Although we still require continuity of temperature at the edge of each cell, the northward extent of the Hadley circulation, ϕ_{H+}, will no longer have the same magnitude as the southward extent, $-\phi_{H-}$. Moreover, the 'equal area' argument must now be applied separately to the northern and southern cells. Recall that in the symmetric case, the requirement of continuity at ϕ_H and the requirement of no net heating (i.e., 'equal area') served to determine both ϕ_H and $\bar{\Theta}(0)$, the temperature at the latitude separating the northern and southern cells – which for the symmetric case is the equator. In the present case, this separating latitude can no longer be the equator. If we choose this latitude to be some arbitrary value, ϕ_1, then the application of temperature continuity and 'equal area' for the northern cell will lead to a value of $\bar{\Theta}(\phi_1)$ that will, in general, be different from the value obtained by application of these same constraints to the southern cell. In order to come out with a unique value for ϕ_1 we must allow ϕ_1 to be a variable to be determined.

The solution is now no longer obtainable analytically, and must be determined numerically. This is easily done with any straightforward search routine. Here we will merely present a few of the results. In

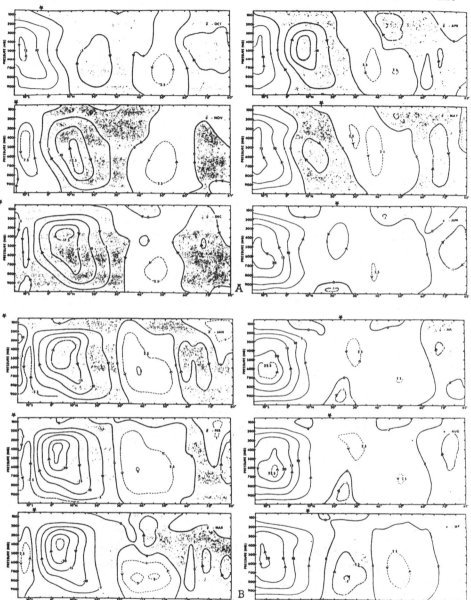

Figure 7.12: Streamlines of the mean meridional circulation for each month. The isolines give the total transport of mass northward below the level considered. Units, 10^{13} gs^{-1} (from Oort and Rasmussen, 1970).

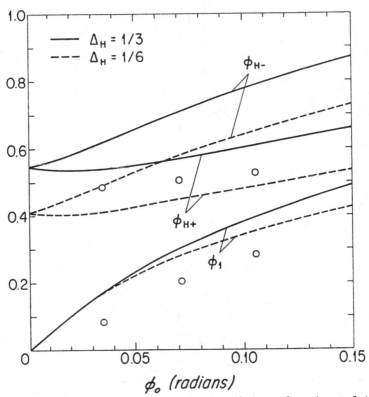

Figure 7.13: The quantities ϕ_{H+}, ϕ_{H-}, and ϕ_1 as functions of ϕ_0 (see text for definitions). Note that 1° of latitude $\cong$ 0.0175 radians (from Lindzen and Hou, 1988).

Figure 7.13 we show how ϕ_{H+}, ϕ_{H-}, and ϕ_1 vary with ϕ_0 for $\Delta_H = 1/3$ (corresponding to a pole–equator temperature difference in $\bar{\Theta}_E$ of about 100°C) and for $\Delta_H = 1/6$. The latter case corresponds to the atmosphere being thermally forced by the surface temperature, and is probably more appropriate for comparisons with observations. For either choice, we see that ϕ_1 goes to fairly large values for small values of ϕ_0. At the same time, ϕ_{H-} also grows to large values while ϕ_{H+} and ϕ_1 asymptotically approach each other – consistent with the northern cell becoming negligible in northern summer. Figure 7.14 shows $\bar{\Theta}$

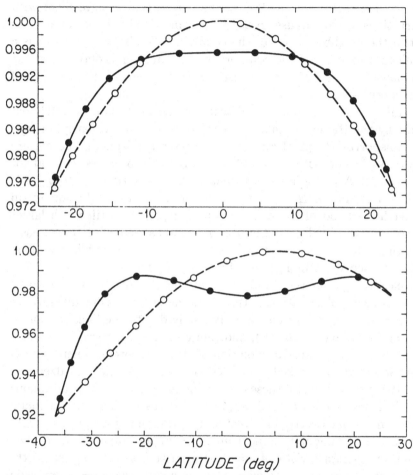

Figure 7.14: $\bar{\Theta}_E/\Theta_0$ (open circles) and $\bar{\Theta}/\Theta_0$ (filled circles) as functions of ϕ obtained with the simple 'equal area' model with $\Delta_H = 1/6$. The upper panel corresponds to $\phi_0 = 0$; the lower panel corresponds to $\phi_0 = 6°$ (from Lindzen and Hou, 1988).

and $\bar{\Theta}_E$ versus latitude for $\phi_0 = 0$ and $\phi_0 = 6°$. We see very clearly the great enlargement and intensification of the southern cell and the corresponding reduction of the northern cell that accompanies the small northward excursion of ϕ_0. We see, moreover, that in agreement with observations at tropopause levels $\bar{\Theta}$ is symmetric about the equator (at least in the neighbourhood of the equator). We also see that $\bar{\Theta}$ has a significant minimum at the equator; such a minimum is observed at the tropopause, but is *not* characteristic of Θ averaged over the depth of the troposphere.

While the simple 'equal area' argument seems to appropriately explain why the Hadley circulation usually consists in primarily a single cell transporting tropical air into the winter hemisphere, the picture it leads to is not without problems. Figure 7.15 shows $u(H, \phi)$ for $\phi_0 = .1$ and $\Delta_H = 1/6$. Consistent with observations, $u(H, \phi_{H+})$ is much weaker than $u(H, \phi_{H-})$ and u is symmetric about the equator in the neighbourhood of the equator, but $u(H, \phi_{H-})$ is still much larger than the observed value, and now $u(H, 0)$ indicates much stronger easterlies than are ever observed. Further difficulties emerge when we look at the surface wind in Figure 7.16. We see that there is now a low level easterly jet on the winter side of the equator; this is, in fact, consistent with observations. However, the surface wind magnitudes (for $\phi_0 = 6°$) are now excessive (only partly due to the linearization of the drag boundary condition), and, more ominously, there are surface westerlies at the equator in violation of Hide's theorem. There are exercises where you are asked to discuss these discrepancies. Lindzen and Hou (1988) show that all these discrepancies disappear in a continuous numerical model with a small amount of viscosity. The discrepancies arise from the one overtly incorrect assumption in the simple approach: namely, that the angular momentum on the upper branch of the Hadley circulation is characteristic of $\phi = \phi_1$ (the latitude separating the northern and southern cells). As we saw in connection with the symmetric Hadley circulation (i.e., $\phi_1 = 0$), the angular momentum in the upper branch was actually characteristic of the entire ascending region. This was not such a significant issue in the symmetric case because the vertical velocity was a maximum at $\phi = \phi_1 = \phi_0 = 0°$. However, when $\phi_0 \neq 0$, then the maximum ascent no longer occurs at ϕ_1; rather it occurs near ϕ_0 where the characteristic angular momentum differs

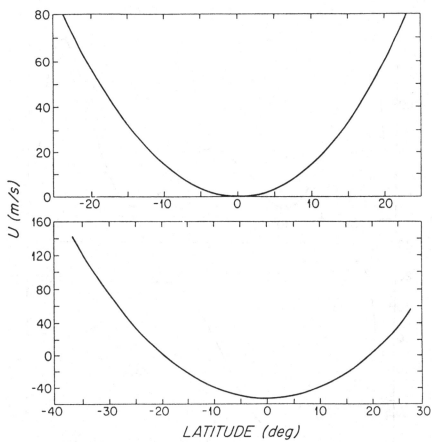

Figure 7.15: Same as Figure 7.14, but for $u(H, \phi)$ (from Lindzen and Hou, 1988).

greatly from that at ϕ_1. It should also be mentioned that in the continuous models, the temperature minimum at the equator is substantially less evident.

The above discussion leads to only modest changes in the five points mentioned in Section 7.3.2. Item 5 is largely taken care of when one recognizes that the Hadley circulation resulting from averaging winter and summer circulations is much larger than the circulation produced by

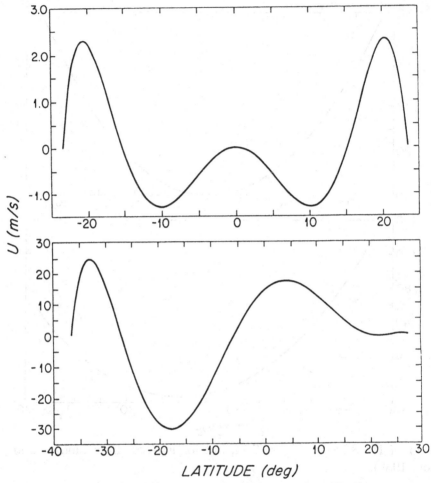

Figure 7.16: Same as Figure 7.14, but for $u(0, \phi)$.

equinoctial forcing. Eddies are probably still needed for the following:

1. to diminish the strength of the jet stream in winter, and, relatedly, to maintain surface winds in middle and high latitudes; and

2. to carry heat between the tropics and the poles.

Lindzen and Hou (1988) stress that Hadley circulations mainly transport angular momentum into the winter hemisphere. Thus, to the extent that eddies are due to the instability of the jet, eddy transports are likely to be mostly present in the winter hemisphere.

In this chapter we have seen how studying the symmetric circulation can tell us quite a lot about the real general circulation – even though a pure symmetric circulation is never observed. In the remainder of this volume, we will focus on the nature of the various eddies. Our view is that eddies are internal waves interacting with the 'mean flow'. Forced waves lose energy to the mean flow, while unstable waves gain energy at the expense of the mean flow.

Exercises

7.1 Calculate u_M (the distribution of u associated with constant M in Equation 7.22) as a function of ϕ for $u_M(0) = 0$.

7.2 Hide's theorem states that in a symmetric circulation a maximum in angular momentum can only exist at the surface in a region of surface easterlies. Moreover, the absolute maximum of angular momentum will be $M = \Omega a^2$, the value associated with zero relative flow at the equator. Using a similar derivation, what can be said about the existence of a minimum in angular momentum in a symmetric circulation?

7.3 Will the simplified approach to calculating zonal wind given by Equations 7.29 and 7.46 necessarily yield adherence to Hide's theorem?

7.4 Derive the equations for the simple model of asymmetric Hadley circulation.

7.5 Try to incorporate the possibility that the angular momentum in the upper, outward flow in the Hadley cells is characteristic of a broad upward flowing region as in Figure 7.6.

Chapter 8

Internal gravity waves, 1

Supplemental reading:

Holton (1979), sections 7.1–7.4

Houghton (1977), sections 8.1–8.3

Lindzen (1973)

Atmospheric waves (eddies) are important in their own right as major components of the total circulation. They are also major transporters of energy and momentum. For a medium to propagate a disturbance as a wave there must be a restoring 'force', and in the atmosphere, this arises, primarily, from two sources: conservation of potential temperature in the presence of positive static stability and from the conservation of potential vorticity in the presence of a mean gradient of potential vorticity. The latter will be treated later; it leads to what are known as Rossby waves. The former leads to internal gravity waves (and surface gravity waves as well). Internal gravity waves are simpler to understand and clearly manifest the various ways in which waves interact with the mean state. For reasons which will soon become clear internal gravity waves are not a dominant part of the midlatitude tropospheric circulation (though they are important). Nonetheless, we will study them in some detail – as prototype atmospheric waves. As a bonus, the theory we develop will be sufficient to allow us to understand atmospheric tides, upper atmosphere turbulence, and the quasi-biennial oscillation of the stratosphere. We will also use our results to speculate on the circulation of Venus. The mathematical apparatus needed will, with one exception, not go beyond understanding the simple harmonic oscillator equation. The exception is that I will use elementary WKB theory (without turning points). Try to familiarize yourself with this device, though it will be briefly sketched in Chapter 10.

8.1   Some general remarks on waves

A wave propagating in the x,z-plane will be characterized by functional dependence of the form

$$\cos(\sigma t - kx - \ell z + \phi).$$

We will refer to $k\hat{i} + \ell\hat{k}$ as the wave vector; the period $\tau = 2\pi/\sigma$; the horizontal wavelength $= 2\pi/k$; the vertical wavelength $= 2\pi/\ell$. Phase velocity is given by

$$\frac{\sigma}{k}\hat{i} + \frac{\sigma}{\ell}\hat{k},$$

while group velocity is given by

$$\frac{\partial\sigma}{\partial k}\hat{i} + \frac{\partial\sigma}{\partial\ell}\hat{k}.$$

When phase velocity and group velocity are the same we refer to the wave as non-dispersive; otherwise it is dispersive; that is, different wavelengths will have different phase speeds and a packet will disperse.

8.1.1   Group and signal velocity

The role of the group velocity in this matter is made clear by the following simple argument. Let us restrict ourselves to a signal $f(t, x)$, where

$$f(t, 0) = C(t)e^{i\omega_0 t}.$$

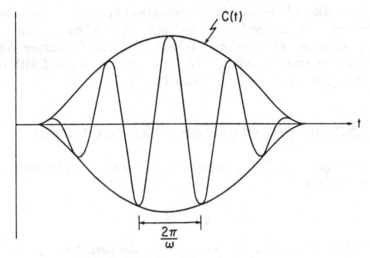

Figure 8.1:   Modulated carrier wave.

We may Fourier expand $C(t)$:

$$C(t) = \int_{-\infty}^{\infty} B(\omega)e^{i\omega t}d\omega,$$

and then

$$f(t,0) = \int_{-\infty}^{\infty} B(\omega)e^{i(\omega + \omega_0)t}d\omega.$$

Away from $x = 0$

$$f(t,x) = \int_{-\infty}^{\infty} B(\omega)e^{i[(\omega + \omega_0)t - k(\omega + \omega_0)x]}d\omega$$

(N.B. $k(\omega + \omega_0)$ means, in this instance, that k is a function of $\omega + \omega_0$).

Now

$$k(\omega + \omega_0) = \left(k\,(\omega_0) + \frac{dk}{d\omega}\right)_{\omega_0} \omega + \ \ldots$$

and

$$f(t,x) = \int_{-\infty}^{\infty} B(\omega)e^{i(\omega_0 t - k(\omega_0)x)}e^{i(\omega t - \frac{dk}{d\omega}\omega x + \cdots)}\,d\omega.$$

Assume $B(\omega)$ is sufficiently band-limited so that the first two terms in the Taylor expansion of k are sufficient. Then

$$\begin{aligned}
f(t,x) &= e^{i(\omega_0 t - k(\omega_0)x)}\int_{-\infty}^{\infty} B(\omega)e^{i\omega(t - \frac{dk}{d\omega}x)}\,d\omega \\
&= e^{i(\omega_0 t - k(\omega_0)x)}C\left(t - \frac{dk}{d\omega}x\right).
\end{aligned}$$

We observe that the information (contained in C) travels with the group velocity

$$c_G = \left(\frac{dk}{d\omega}\right)^{-1} = \frac{d\omega}{dk}.$$

8.2 Heuristic theory (no rotation)

In studying atmospheric thermodynamics you have seen that a neutrally buoyant blob in a stably stratified fluid will oscillate up and down with a frequency N. Applying this to the configuration in Figure 8.2, we get

$$\frac{d^2\delta s}{dt^2} = -N^2\delta s,$$

where

$$N^2 = \underbrace{\frac{g}{T_0}\left(\frac{dT_0}{dz} + \frac{g}{c_p}\right)}_{\text{static stability}} \quad \left(-\frac{g}{\rho_0}\frac{d\rho_0}{dz}\text{ in a Boussinesq fluid}\right).$$

The restoring force (per unit mass), $F_b = -N^2\delta s$, is directed vertically.

Now consider the following situation where a corrugated sheet is pulled horizontally at a speed c through a stratified fluid (*viz.*, Figure 8.3). Wave motions will be excited in the fluid above the plate. For

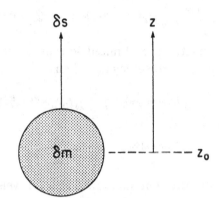

Figure 8.2:   Blob in stratified fluid.

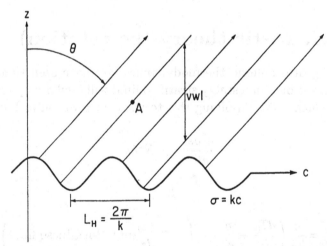

Figure 8.3:   Corrugated lower surface moving through a fluid.

our oscillating blob we used '$F = ma'$. Normally, we cannot use this for fluid flows because of the pressure force. However, in a wave field there will be lines of constant phase. Along these lines the pressure perturbation will be constant and hence pressure gradients along such lines will be zero, and the acceleration of the fluid along such lines will, indeed, be given by $F = ma$.

Assume such lines make an angle Θ with respect to the vertical. The projection of the buoyancy force is then

$$F = -N^2 \delta z \cos \Theta.$$

Also

$$\delta z = \delta s \cos \Theta$$

so

$$F = -N^2 \cos^2 \Theta \, \delta s$$

and

$$\frac{d^2 \delta s}{dt^2} = -N^2 \cos^2 \Theta \, \delta s.$$

Thus

$$\sigma^2 = N^2 \cos^2 \Theta = k^2 c^2;$$

that is, kc determines Θ.

Θ is also related to the ratio of horizontal and vertical wavelengths:

$$\tan \Theta \;=\; L_H / L_v = \ell / k$$

$$\tan^2 \Theta \;=\; \frac{\ell^2}{k^2} = \frac{1 - \cos^2 \Theta}{\cos^2 \Theta} = \frac{1 - \frac{k^2 c^2}{N^2}}{\frac{k^2 c^2}{N^2}},$$

which is, in fact, our dispersion relation.

$$\ell^2 = \left(\frac{N^2}{k^2 c^2} - 1\right) k^2 = \left(\frac{N^2}{\sigma^2} - 1\right) k^2. \tag{8.1}$$

We see, immediately, that vertical propagation requires that $\sigma^2 < N^2$. When $\sigma^2 > N^2$, the buoyancy force is inadequate to maintain an oscillation and the perturbation decays with height. Equation 8.1 may be solved for σ:

$$\sigma = \pm \frac{Nk}{(k^2 + \ell^2)^{1/2}} \tag{8.2}$$

Now

$$c_{p_x} = \frac{\sigma}{k} = \pm \frac{N}{(k^2 + \ell^2)^{1/2}} \tag{8.3}$$

$$c_{p_z} = \frac{\sigma}{\ell} = \pm \frac{Nk/\ell}{(k^2 + \ell^2)^{1/2}} \tag{8.4}$$

$$c_{g_x} = \frac{\partial \sigma}{\partial k} = \pm \frac{N\ell^2}{(k^2 + \ell^2)^{3/2}} \tag{8.5}$$

$$c_{g_z} = \frac{\partial \sigma}{\partial \ell} = \pm \frac{-Nk\ell}{(k^2 + \ell^2)^{3/2}}. \tag{8.6}$$

Thus c_{p_x} and c_{g_z} are of the same sign, while c_{p_z} and c_{g_z} are of opposite sign. What this means is easily seen from Figure 8.3. Since the wave is forced by the moving corrugated plate, the motions must angle to the right of the vertical so that the moving corrugation is pushing the fluid ($\overline{p'w'}$ is positive). (Equation 8.2 permits Θ to be positive or negative, but negative values would correspond to the fluid pushing the plate.) But then phase line A, which is moving to the right, is also seen by an observer at a fixed x as moving downward. Thus for an internal gravity wave, upward wave propagation is associated with downward phase propagation[1].

Before moving beyond this heuristic treatment, two points should be mentioned.

1. We have already noted that vertical propagation ceases, for positive k^2, when σ^2 exceeds N^2. (For the atmosphere $2\pi/N \sim$ 5 min.) It is also worth noting what happens as $\sigma \to 0$. The vertical wavelength $(2\pi/\ell) \to 0$, as does c_{g_z}. We may anticipate that in this limit any damping will effectively prevent vertical propagation. Why?

2. The excitation of gravity waves by a moving corrugated plate does not seem terribly relevant to either the atmosphere or the ocean. So we should see what is needed more generally. What is needed is anything which will, as seen by an observer moving with the fluid, move height surfaces up and down. Thus, rather than

---

[1] This is not always true if the fluid is moving relative to the observer. The reader is urged to examine this possibility.

move the corrugations through the fluid, it will suffice for the fluid to move past fixed corrugations – or mountains for that matter. Similarly, a heat source moving relative to the fluid will displace height surfaces and excite gravity waves. The daily variations of solar insolation act this way. Other more subtle excitations of gravity waves arise from fluid instabilities, collapse of fronts, squalls, and so forth.

The above, heuristic, analysis tells us much about gravity waves – and is simpler and more physical than the direct application of the equations. However, it is restricted to vertical wavelengths much shorter than the fluid's scale height; it does not include rotation, friction, or the possibility that the unperturbed basic state might have a spatially variable flow. To extend the heuristic approach to such increasingly complicated situations becomes almost impossible. It is in these circumstances that the equations of motion come into their own.

8.3 Linearization

Implicit in the above was that the gravity waves were small perturbations on the unperturbed basic state. What does this involve in the context of the equations of motion? Consider the equation of x-momentum on a non-rotating plane for an inviscid fluid:

$$\frac{\partial u}{\partial t} + u\frac{\partial u}{\partial x} + v\frac{\partial u}{\partial y} + w\frac{\partial u}{\partial z} = -\frac{1}{\rho}\frac{\partial p}{\partial x}. \tag{8.7}$$

In the absence of wave perturbations assume a solution of the following form:

$$
\begin{aligned}
u &= u_0(y, z) \\
p &= p_0(y, z) \\
\rho &= \rho_0(y, z) \\
v_0 &= 0 \\
w_0 &= 0.
\end{aligned}
$$

Equation 8.7 is automatically satisfied. Now, add to the basic state, perturbations u', v', w', ρ', p'. Equation 8.7 becomes

$$\frac{\partial u'}{\partial t} + u_0\frac{\partial u'}{\partial x} \quad + \quad u'\frac{\partial u'}{\partial x} + v'\frac{\partial u_0}{\partial y} + v'\frac{\partial u'}{\partial y}$$

$$+ \quad w'\frac{\partial u_0}{\partial z} + w'\frac{\partial u'}{\partial z} = -\frac{1}{(\rho_0 + \rho')}\frac{\partial p'}{\partial x}. \qquad (8.8)$$

Terms involving the basic state alone cancelled since the basic state must, itself, be a solution. Linearization is possible when the perturbation is so small that terms quadratic in the perturbation are much smaller than terms linear in the perturbation. The linearization of Equation 8.8 is

$$\frac{\partial u'}{\partial t} + u_0\frac{\partial u'}{\partial x} + v'\frac{\partial u_0}{\partial y} + w'\frac{\partial u_0}{\partial z} = -\frac{1}{\rho_0}\frac{\partial p'}{\partial x}. \qquad (8.9)$$

It is sometimes stated that linearization requires $|u'| \ll |u_0|$, but clearly this is too restrictive. The precise assessment of the validity of linearization depends on the particular problem being solved. For our treatment of gravity waves we will always assume the waves to be linearizable perturbations.

Before proceeding to explicit solutions we will prove a pair of theorems which are at the heart of wave–mean flow interactions. Although we will not be using these theorems immediately, I would like to present them early so that there will be time to think about them. I will also discuss the relation between energy flux and the direction of energy propagation.

8.4   Eliassen-Palm theorems

Let us assume that rotation may be ignored. Let us also ignore viscosity and thermal conductivity. Finally, let us restrict ourselves to basic flows where $v_0 = w_0 = 0$ and $u_0 = u_0(z)$. Also, let us include thermal forcing of the form

$$J = \tilde{J}(z)e^{ik(x-ct)}$$

and seek solutions with the same x and t dependence. The equation for x-momentum becomes

$$\rho_0 \left(\frac{\partial u'}{\partial t} + u_0 \frac{\partial u'}{\partial x} + w' \frac{\partial u_0}{\partial z} \right) + \frac{\partial p'}{\partial x} = 0$$

or

$$\rho_0(u_0 - c)\frac{\partial u'}{\partial x} + \rho_0 w' \frac{du_0}{dz} + \frac{\partial p'}{\partial x} = 0. \tag{8.10}$$

Similarly for w',

$$\rho_0(u_0 - c)\frac{\partial w'}{\partial x} + \rho' g + \frac{\partial p'}{\partial z} = 0. \tag{8.11}$$

Continuity yields

$$\frac{\partial u'}{\partial x} + \frac{\partial w'}{\partial z} + \frac{1}{\rho_0}\left((u_0 - c)\frac{\partial \rho'}{\partial x} + w'\frac{d\rho_0}{dz} \right) = 0. \tag{8.12}$$

From the energy equation (expressed in terms of p and ρ instead of T and ρ; do the transformations yourself)

$$(u_0 - c)\frac{\partial p'}{\partial x} + w'\frac{dp_0}{dz} = \gamma g H \left((u_0 - c)\frac{\partial \rho'}{\partial x} + w'\frac{d\rho_0}{dz} \right) + (\gamma - 1)\rho_0 J. \tag{8.13}$$

(Remember that $\gamma = c_p/c_v$.) Now multiply Equation 8.10 by $(\rho_0(u_0 - c)u' + p')$ and average over x:

$$(\rho_0(u_0 - c)u' + p') \quad \cdot \quad \frac{\partial}{\partial x}(\rho_0(u_0 - c)u' + p')$$

$$+ \quad \rho_0 \frac{du_0}{dz}(\rho_0(u_0 - c)u'w' + p'w') = 0.$$

The first term vanishes when averaged over a wavelength in the x-direction while the second term yields

$$\overline{p'w'} = -\rho_0(u_0 - c)\overline{u'w'}. \tag{8.14}$$

Equation 8.14 is Eliassen and Palm's first theorem. $\overline{p'w'}$ is the vertical energy flux associated with the wave (this is not quite true – but its sign is the sign of wave propagation; we will further discuss this later in this chapter), while $\rho_0 \overline{u'w'}$ is the vertical flux of momentum carried by the wave (Reynold's stress). This theorem tells us that the momentum flux is such that if deposited in the mean flow it will bring u_0 towards c. Stated differently, an upward propagating wave carries westerly momentum if $c > u_0$ and easterly momentum if $c < u_0$!

Their second theorem, which tells how $\rho_0 \overline{u'w'}$ varies with height, is harder to prove.

Let

$$ik(u_0 - c)\zeta = w = (u_0 - c)\frac{\partial \zeta}{\partial x} \tag{8.15}$$

(this simply defines the vertical displacement, ζ, for small perturbations). Substituting Equation 8.15 into Equation 8.13 yields

$$\frac{p'}{\gamma g H} + \frac{\rho_0}{g}N^2\zeta = \rho' + \frac{(\gamma - 1)}{\gamma g H}\frac{\rho_0 J}{ik(u_0 - c)}. \tag{8.16}$$

Substituting Equation 8.16 into Equation 8.11 yields

$$\rho_0(u_0 - c)\frac{\partial w'}{\partial x} + \frac{p'}{\gamma H} + \rho_0 N^2\zeta - \frac{(\gamma - 1)}{\gamma H}\frac{\rho_0 J}{ik(u_0 - c)} + \frac{\partial p'}{\partial z} = 0. \tag{8.17}$$

Substituting Equation 8.13 into Equation 8.12 yields

$$\frac{\partial u'}{\partial x} + \frac{\partial w'}{\partial z} + \frac{1}{\rho_0 \gamma g H}(u_0 - c)\frac{\partial p'}{\partial x} - \frac{1}{\gamma H}w' - \frac{(\gamma - 1)}{\gamma g H}J = 0. \tag{8.18}$$

Now multiply Equation 8.10 by u', Equation 8.17 by w', and Equation 8.18 by p' and sum them. Equation 8.10 multiplied by u' yields

$$\frac{\partial}{\partial x}\left(\frac{\rho_0(u_0 - c)u'^2}{2}\right) + \rho_0\frac{du_0}{dz}u'w' + u'\frac{\partial p}{\partial x} = 0. \tag{8.19}$$

Equation 8.17 multiplied by w' yields

$$\frac{\partial}{\partial x}\left(\frac{\rho_0(u_0 - c)w'^2}{2}\right) + \frac{w'p'}{\gamma H} + \frac{\partial}{\partial x}\left(\frac{\rho_0 N^2(u_0 - c)\zeta^2}{2}\right)$$
$$- \frac{(\gamma - 1)\rho_0\zeta J}{\gamma H} + w'\frac{\partial p'}{\partial z} = 0, \tag{8.20}$$

where Equation 8.15 has been used to define ζ.

Equation 8.18 multiplied by p' yields

$$p'\frac{\partial u'}{\partial x} + p'\frac{\partial w'}{\partial z} + \frac{\partial}{\partial x}\left(\frac{1}{2}\frac{(u_0 - c)}{\rho_0\gamma gH}p'^2\right) - \frac{p'w'}{\gamma H} - \frac{\kappa}{gH}p'J = 0. \quad (8.21)$$

Adding Equations 8.19, 8.20, and 8.21 yields

$$\frac{\partial}{\partial x}\left\{\frac{1}{2}\rho_0(u_0 - c)u'^2 + \frac{1}{2}\rho_0(u_0 - c)w'^2 + \frac{1}{2}\rho_0 N^2(u_0 - c)\zeta^2\right.$$
$$\left. + \frac{1}{2}\frac{(u_0 - c)}{\rho_0\gamma gH}p'^2 + p'u'\right\}$$
$$+\frac{\partial}{\partial z}(p'w') + \rho_0\frac{du_0}{dz}u'w' - \frac{\kappa\rho_0\zeta J}{H} - \frac{\kappa}{gH}p'J = 0 \quad (8.22)$$

(where $\kappa = \frac{\gamma - 1}{\gamma}$).

The last two terms can be rewritten

$$-\kappa\rho_0\left(\frac{\zeta}{H} + \frac{p'}{p_0}\right)J.$$

Averaging with respect to x (assuming periodicity):

$$\frac{\partial}{\partial z}(\overline{w'p'}) = -\frac{d\bar{u}}{dz}\rho_0\overline{u'w'} + \kappa\rho_0\overline{\left(\frac{\zeta}{H} + \frac{p'}{p_0}\right)J}. \quad (8.23)$$

Finally, we substitute Equation 8.14 in Equation 8.23 to obtain one version of Eliassen and Palm's second theorem:

$$\frac{\partial}{\partial z}(\rho_0\overline{u'w'}) = -\frac{\kappa\rho_0}{(u_0 - c)}\overline{DJ}, \quad (8.24)$$

where

$$D = \frac{\zeta}{H} + \frac{p'}{p_0}.$$

Eliassen and Palm (1961) developed their theorem for the case where $J = 0$. In that case

$$\frac{\partial}{\partial z}(\rho_0\overline{u'w'}) = 0. \quad (8.25)$$

Notice that theorem 1 (Equation 8.14) tells us that the sign of $\rho_0 \overline{u'w'}$ is independent of $\frac{du_0}{dz}$, while theorem 2 (Equation 8.25) tells us that in the absence of

(i)   damping,

(ii) local thermal forcing, and

(iii) critical levels (where $u_0 - c = 0$)

no momentum flux is deposited or extracted from the basic flow. Contrast Equations 8.14 and 8.25 with what one gets assuming Reynold's stresses are due to locally generated turbulence (where eddy diffusion is down-gradient), and consider the fact that in most of the atmosphere, eddies are, in fact, waves.

From Equation 8.25 and Equation 8.14 we also see that the quantity $\overline{p'w'}/(u_0-c)$ and not $\overline{p'w'}$ is conserved. The former is sometimes referred to as wave action.

8.4.1 'Moving flame effect' and the super-rotation of Venus' atmosphere

The role of the right-hand side of Equation 8.24 is of some interest. It is clear that $\overline{DJ} \neq 0$, at least in a situation which allows the radiation of waves to infinity. Consider thermal forcing in some layer (as shown in Figure 8.4). Above J there will be a momentum flux which must come from some place and cannot come from the region below J; hence, it must come from the thermal forcing region. Moreover, the flux divergence within the heating region must be such as to accelerate the fluid within the heating region in a direction opposite to c. This mechanism has sometimes been called the 'moving flame effect,' and has been suggested as the mechanism responsible for maintaining an observed 100 m/s zonal flow in Venusian cloud layer, the thermal forcing being due to the absorption of sunlight at the cloud top. Since this flow is in the direction of Venus' rotation, it is referred to as 'super-rotation'.

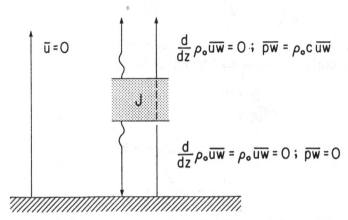

Figure 8.4: Fluxes associated with a layer of thermal forcing.

8.5 Energy flux

We have already noted that $\overline{p'w'}$ is not the complete expression for the wave flux of energy. It is merely the contribution of the pressure-work term to the flux. In addition, there is the advection by the wave fields of the kinetic energy of the mean flow, $\rho_0 \overline{u'w'} U$ (Show this.). Thus, the full expression for the energy flux is

$$F_E = \overline{p'w'} + \rho_0 \overline{u'w'} U. \tag{8.26}$$

Note that now $dF_E/dz = 0$ under the conditions for which non-interaction holds. On the other hand, F_E is arbitrary up to a Galilean transformation, and hence may no longer be associated with the direction of wave propagation.

The above difficulty does not depend on U having shear, so we will restrict ourselves to the case where $dU/dz = 0$. We will also ignore time variations in ρ_0 – which is acceptable for a Boussinesq fluid. Finally, we will assume the presence of a slight amount of damping which will lead to some absorption of wave fluxes and the consequent modification of the basic state:

$$\rho_0 \frac{\partial U}{\partial t} = -\frac{d}{dz} \rho_0 \overline{u'w'} \tag{8.27}$$

and

$$\rho_0 \frac{\partial}{\partial t}\left(\frac{U^2}{2} + Q\right) = -\frac{d}{dz}(\overline{p'w'} + \rho_0\overline{u'w'}U), \qquad (8.28)$$

where Q is a measure of the fluid's heat or internal energy.

Now multiply Equation 8.27 by U,

$$\rho_0 \frac{\partial}{\partial t}\left(\frac{U^2}{2}\right) = -U\frac{d}{dz}\rho_0\overline{u'w'}, \qquad (8.29)$$

and subtract Equation 8.29 from Equation 8.28,

$$\rho_0 \frac{\partial}{\partial t}Q = -\frac{d}{dz}\overline{p'w'}. \qquad (8.30)$$

From Equation 8.29, we see that the second term in Equation 8.26 is associated with the alteration of the kinetic energy of the basic state (associated with wave absorption), and that the choice of a Galilean frame can lead to either a decrease or an increase of the mean kinetic energy (Why?). From Equation 8.30 we see that the convergence of $\overline{p'w'}$ is, on the other hand, associated with mean heating. Since we do not wish the absorption of a wave to cool a fluid, we are forced to identify the direction of $\overline{p'w'}$ with the direction of wave propagation. (Recall that the wave is attenuated in the direction of propagation.)

8.6   A remark about 'eddies'

As we proceed with our discussion of eddies, it will become easy to lose track of where we are going. Recall this course has a number of aims:

(i)   to familiarize you with the foundations and methodologies of dynamics;

(ii) to use this tool to account for some of the observed motion systems of the atmosphere and oceans; and

(iii) to examine the roles of these systems in the 'general circulation'.

You can be reasonably assured that our approach to (ii) will *not* be very systematic. Even with item (iii), it will not be easy to provide a straightforward treatment – at least partly because a complete answer is not yet available!

In our treatment of the symmetric circulation our approach was to calculate a symmetric circulation and see to what extent it accounted for observations. Our assumption was that the degree to which symmetric models failed pointed the way to the role of eddies. We will follow a similarly indirect path in studying eddies and their rôles. The difficulty here is that there are many kinds of eddies! Broadly speaking, we have gravity and Rossby waves – but these may be forced, or they may be free 'drum head' oscillations, or they may even be instabilities of the basic flow. We shall not, of course, study these various eddies at random. We investigate various waves because they seem suitable to particular phenomena. However, there always remains a strong element of 'seeing what happens' which should be neither forgotten nor underestimated.

8.7   Mathematical treatment

We shall now return to the problem considered heuristically earlier in this chapter in order to deal with it in a more formal manner. This approach will confirm and extend the results obtained heuristically. The new results are important in their own right; they will also permit us to introduce terminology and concepts which are essential to the discussion of atmospheric tides in Chapter 9. Our equations will be the Boussinesq equations

$$\rho_0 \frac{\partial u'}{\partial t} = -\frac{\partial p'}{\partial x} \tag{8.31}$$

$$\rho_0 \frac{\partial w'}{\partial t} = -\frac{\partial p'}{\partial z} - g\rho' \tag{8.32}$$

$$\frac{\partial u'}{\partial x} + \frac{\partial w'}{\partial z} = 0 \tag{8.33}$$

$$\frac{\partial \rho'}{\partial t} + w' \frac{d\rho_0}{dz} = 0, \tag{8.34}$$

where ρ_0 and $\frac{d\rho_0}{dz}$ are treated as constants.

We seek solutions of the form $e^{i(kx-\sigma t)}$:

$$-\rho_0 i\sigma u + ikp = 0 \tag{8.35}$$

$$-\rho_0 i\sigma w + p_z + g\rho = 0 \tag{8.36}$$

$$-i\sigma\rho + w\rho_{0z} = 0 \tag{8.37}$$

$$iku + w_z = 0. \tag{8.38}$$

(For convenience we have dropped primes on perturbation quantities.) Equations 8.35 and 8.38 imply

$$ik\left(\frac{k}{\sigma}\frac{p}{\rho_0}\right) + w_z = 0. \tag{8.39}$$

Equations 8.36 and 8.37 imply

$$-\rho_0 i\sigma w + p_z + g\left(\frac{\rho_{0z}}{i\sigma}w\right) = 0, \tag{8.40}$$

and eliminating p between Equations 8.39 and 8.40 implies

$$w_{zz} + \left\{\left(-\frac{g\rho_{0z}}{\sigma^2\rho_0} - 1\right)k^2\right\}w = 0, \tag{8.41}$$

or

$$w_{zz} + \left\{\left(\frac{N^2}{\sigma^2} - 1\right)k^2\right\}w = 0. \tag{8.42}$$

Note that if Equation 8.36 is replaced by the hydrostatic relation

$$p_z + g\rho = 0, \tag{8.43}$$

then Equation 8.42 becomes

$$w_{zz} + \left\{\frac{N^2 k^2}{\sigma^2}\right\}w = 0;$$

that is, hydrostaticity is okay if $N^2/\sigma^2 \gg 1$. This is essentially the same result we obtained in Chapter 6 by means of scaling arguments.

If we assume solutions of the form $e^{i\ell z}$, Equation 8.42 is identical to our heuristic result.

Equation 8.41 allows us to examine the relation between internal and surface waves. Note first, that Equation 8.42 has no homogeneous solution for an unbounded fluid that satisfies either the radiation condition or boundedness as $z \to \infty$. (Such homogeneous solutions will, however, be possible when we allow such features as compressibility and height variable basic stratification.) For Equation 8.42 to have homogeneous solutions (free oscillations) there must be a bounding upper surface. If this is a free surface (such as the ocean surface; the atmosphere doesn't have such a surface), then the appropriate upper boundary condition is $dp/dt = 0$. Linearizing this condition yields

$$\frac{\partial p'}{\partial t} + w'\frac{dp_0}{dz} = \frac{\partial p'}{\partial t} - w'g\rho_0 = 0 \quad \text{at } z = \text{H}. \tag{8.44}$$

Using Equations 8.35 and 8.38, Equation 8.44 becomes

$$w_z = g\frac{k^2}{\sigma^2}w \text{ at } z = \text{H}. \tag{8.45}$$

Solving Equation 8.42 subject to Equation 8.45 and the condition

$$w = 0 \text{ at } z = 0 \tag{8.46}$$

leads to our free oscillations. Equation 8.42 will have solutions of the form

$$w = \sinh \mu z,$$

where

$$\mu^2 = \left(1 - \frac{N^2}{\sigma^2}\right)k^2, \tag{8.47}$$

or

$$w = \sin \lambda z,$$

where

$$\lambda^2 = \left(\frac{N^2}{\sigma^2} - 1\right)k^2, \tag{8.48}$$

depending on whether σ^2 is greater or less than N^2. Inserting Equations 8.47 and 8.48 into Equation 8.45 yields

$$\tanh \mu H = \frac{\mu}{g} \frac{\sigma^2}{k^2} \qquad (8.49)$$

and

$$\tan \lambda H = \frac{\lambda}{g} \frac{\sigma^2}{k^2}, \qquad (8.50)$$

respectively.

8.7.1 Shallow water limit and internal modes

In the limit of a shallow fluid ($\mu H \ll 1$ or $\lambda H \ll 1$) both Equation 8.49 and Equation 8.50 reduce to

$$\frac{\sigma^2}{k^2} = gH, \qquad (8.51)$$

which is independent of N^2. (What do the solutions look like?) In addition, Equation 8.50 has an infinite number of solutions. In order to examine the nature of these solutions it suffices to replace Equation 8.45 with

$$w = 0 \quad \text{at } z = H. \qquad (8.52)$$

Then our solutions become

$$w = \sin \lambda z,$$

where

$$\lambda = \left(\frac{N^2}{\sigma^2} - 1 \right)^{1/2} k = \frac{n\pi}{H}, \quad n = 1, 2, \ldots . \qquad (8.53)$$

For simplicity let us use the hydrostatic approximation. Then Equation 8.53 becomes

$$\lambda = \frac{N}{\sigma}k = \frac{n\pi}{H}.$$ (8.54)

8.7.2 Equivalent depth

Solving Equation 8.54 for σ^2/k^2 we get

$$\frac{\sigma^2}{k^2} = \frac{N^2H^2}{n^2\pi^2} \equiv gh.$$ (8.55)

By analogy with Equation 8.51, h is referred to as the equivalent depth of the internal mode (or free oscillation). Similarly, in a forced problem, where we impose σ and k, the relation

$$\frac{\sigma^2}{k^2} = gh$$

defines an equivalent depth for the forced mode. (What happens in a channel where $v = 0$ at $y = 0, L$?) This is not, in general, the equivalent depth of any particular free oscillation. If it is we have resonance. These terms (generalized to a rotating, spherical atmosphere) play a very important role in our discussion of atmospheric tides.

Exercises

In the appendix, we describe a computer model for conveniently studying internal gravity behaviour under a wide variety of circumstances. It proves convenient to use $\log -p$ coordinates in this model. The implementation of this model will add greatly to the value of this and the following two chapters. It will also form a useful tool for many other purposes. The first problem for this chapter allows the reader to become familiar with $\log -p$ coordinates. The notation is that used in this chapter and in the appendix. The remaining problems assume that you have implemented the numerical model, and require its use.

8.1 *Eliassen-Palm theorems in log $-p$ coordinates* In this problem you are going to prove the Eliassen-Palm theorems in $\log -p$ coordinates and deduce the corresponding wave fluxes.

(a) From the nonlinear equations, show that

$$e^{-z^*}\frac{\partial U_0}{\partial t} = -\frac{d}{dz}(e^{-z^*}\overline{uw^*})$$

(assuming $U_0 = \bar{U}$)

(b) Verify

$$\overline{\Phi'w^{*\prime}} = -(U_0 - c_r)\overline{u'w^{*\prime}} - \frac{a}{U_{oz^*}}((U_0 - c_r)\overline{u'u'} + \overline{\Phi'u'}),$$

where $c_r = \text{Real}\{c\}$.

(c) Demonstrate

$$\frac{d}{dz^*}(e^{-z^*}\overline{u'w^{*\prime}}) = \frac{+1}{2k}(F(z^*)\text{Imag}\{\tilde{w}\} + \text{Imag}\{Q^2\}|\tilde{w}|^2).$$

(Hint: show first that

$$\frac{d}{dz^*}(e^{-z^*}\overline{u'w^{*\prime}}) = \frac{-i}{4k}(\tilde{w}\tilde{w}^\dagger_{z^*z^*} - \tilde{w}^\dagger\tilde{w}_{z^*z^*}).)$$

8.2 *Effect of numerical resolution on answers*

Run the model described in the appendix for a wave forced from the lower boundary with the radiation condition imposed at 60 km. For simplicity, take Q constant and corresponding to a vertical wavelength of 5 km. Vary the number of levels from 30 to 500. Discuss the effect of numerical resolution on answers.

You should have noticed, in doing the above, that the numerically obtained amplitude, in addition to increasing exponentially, oscillates with height. This is indicative of partial reflection at the upper boundary. Discuss the cause of this spurious reflection.

The following questions refer to 'suggested runs' using the numerical model in the appendix. The parameters for these runs are given in a table at the end of this problem set.

Table 8.1: Gravity Wave Runs. The notation in this table is that used in the appendix. The units for height are kilometers, for speeds, meters/second, and for temperature, degrees Kelvin. 'L' refers to a lid; 'R' refers to the radiation condition.

Run	Levels	Z_{top}	Boundary Condition, Phase Velocity, Basic State Forcing								
			W_{bot}	Upper	Re(c)	Im(c)	T_0	U_0	A	Z_f	Z_w
1	500	29.5	1	L	30	0	300	10	0	1	1
1a	500	29.5	1	L	30	1	300	10	0	1	1
2	500	29.7	1	L	30	0	300	10	0	1	1
2a	500	29.7	1	L	30	1	300	10	0	1	1
3	500	30.0	1	L	30	0	300	10	0	1	1
3a	500	30.0	1	L	30	1	300	10	0	1	1
4	500	30.5	0	R	30	0	300	10	10	10	1
5	500	30.5	0	R	30	0	300	10	10	11	1
6	500	30.5	0	R	30	0	300	10	10	12	1
7	500	30.5	0	R	30	0	300	10	10	13	1
8	500	30.5	0	R	30	0	300	10	10	14	1
9	500	30.5	0	R	30	0	300	10	10	15	1
10	500	29.7	0	L	30	0	300	10	10	10	1
11	500	29.7	0	L	30	0	300	10	10	12	1
12	500	30.5	0	R	30	0	300	10	10	13	6
13	500	30.5	0	R	30	0	300	10	10	15	6
14	500	30.5	0	R	30	0	300	10	10	0	6
15	500	30.5	0	R	30	0	300	10	10	0	12
16	500	29.7	0	L	30	0	300	10	10	13	6
17	500	29.7	0	L	30	0	300	10	10	15	6

8.3 *Resonances*

Examine waves forced at the lower boundary when a lid is imposed at z_{top}. As z_{top}^* is varied the system manifests resonances. Compare these with analytical results. Discuss both the nature of resonance and the difference between numerical and analytic results. What is the effect of damping? (Suggested runs: 1, 2, 3, 1a, 2a, and 3a.)

8.4 *'Thin' Forcing*

(a) By 'thin' we mean a forcing whose dimensions are small compared to the vertical wavelength. Impose the radiation condition at the 'top' and set $\tilde{w}_{bot} = 0$. Move the forcing up and down and discuss what happens, in particular to the sign of the wave flux $F_e = \exp(-z^*)\overline{\Phi' w^{*'}}$. Is the second

E-P theorem satisfied? Where? (Suggested runs: 4,5,6,7,8, and 9.)

(b) Use 'thin' forcing and impose a lid at the same level you found resonances in (1). Move the forcing up and down as in (a). Discuss the results. (Suggested runs: 10 and 11.)

8.5 'Broad' Forcing

(a) Again, impose the radiation condition at the top and set $\tilde{w}_{bot} = 0$, but now consider a forcing whose dimensions are bigger than the vertical wavelength. Move the forcing up and down. Notice that very little wave emerges from the forcing region until the forcing region is brought into contact with the ground. Then the wave emerging from the forcing region becomes relatively independent of z_w, the half width of the forcing. Why? (Suggested runs: 12, 13, 14, and 15.)

(b) Impose a lid at the level you found resonances in (1). What is different from (a)? Discuss. (Suggested runs: 16 and 17.)

Chapter 9

Atmospheric tides

Supplemental reading:

Chapman and Lindzen (1970)

Lindzen and Chapman (1969)

Lindzen (1979)

Lindzen (1967a)

One of the most straightforward and illuminating applications of internal gravity wave theory is the explanation of the atmosphere's tides. To be sure, the theory has to be expanded to include both the sphericity and rotation of the earth.

By atmospheric tides we generally mean those planetary scale oscillations whose periods are integral fractions of a solar or lunar day (*diurnal* refers to a period of one day, *semidiurnal* refers to a period of half a day, and *terdiurnal* refers to a period of one third of a day). These periods are chosen because we know there is forcing at these periods. Gravitational forcing is precisely known; thermal forcing (due in large measure to the absorption of sunlight by O_3 and water vapor) is known with less precision. Nevertheless, a situation where forcing of known frequency is even reasonably well known is a situation of rare simplicity, and we may plausibly expect that our ability to calculate the observed response to such forcing constitutes a modest test of the utility of theory.

The situation was not always so simple. There follow sections on the history of this problem and on the observations of atmospheric tides. The history also provides a good example of what constitutes the 'scientific method' in an observational science where controlled experiments are not available.

9.1   History and the 'scientific method'

Textbooks in meteorology (and most other sciences) usually treat history (if they treat it at all) as an entertaining diversion from the 'meat' of a subject. I would hardly deny the fact that history is entertaining; however, I also happen to think that history is basic to the subject. In any field where there has been any success at all, one ought to see how significant problems were actually defined and solved (at least to the extent that they were defined and solved). From this point of view, this book actually devotes too little space to history. Thus, the present brief history of the study of atmospheric tides will have to serve as a surrogate for all the omitted histories of other topics we have covered. As such it is a relatively good choice. Just as atmospheric tides constitutes a relatively simple problem in dynamic meteorology, so too the history of this topic (at least until recently) has also been relatively easy to describe. To be sure a professional historian might balk at such a remark, but hopefully, the reader will be more indulgent of an amateur's approach. The history of the study of atmospheric tides provides a particularly good example of the form and pitfalls of the 'scientific method' in an observational science such as meteorology. Controlled experiments are generally out of the question. Instead, one begins with incompletely observed phenomena which are addressed by theoretical explanations. Explanations which go no further than dealing with the partial observations are more nearly simulations than theories, and given human nature, it is usually pretty certain that one will simulate pretty well what has been already observed. A theory should go further – it should offer predictions that go beyond the present observations so that the credibility of the theory can be tested as new observations are made. Quite properly, failure to confirm predictions tends to discredit theories, but confirmation does not as a rule rigorously establish the correctness of a theory; it merely increases our confidence in the theory. The process, in fact, can continue almost indefinitely, though at some point our confidence may seem so well founded that further tests will have a lower priority. The whole process is muddied by the fact (mentioned in Chapter 5) that meteorological data itself frequently is subject to substantial uncertainty. We will see examples of all these factors in the history of atmospheric tides.

In contrast to sea tides, which have been known and described for over two thousand years, atmospheric tides were not observed until the invention of the barometer by Torricelli (*ca.* 1643)[1]. Newton was able to explain the dominance of the lunar semidiurnal component of the sea tide. Briefly, tidal forcing depends not only on the average gravitational force exerted by either the sun or moon, but also on the relative variation of this force over the diameter of the earth. This latter factor gives a substantial advantage to the moon. The dominance of the semidiurnal component arises because, *relative* to the solid earth, the gravitational pull of the sun or moon or any other body simultaneously attracts the portion of the fluid shell directly under it and repels that portion of the envelope opposite it. From the perspective of the earth, both represent outward forces. Thus, in a single period of rotation the fluid envelope is pushed outward twice[2]. Newton already recognized that there ought to be a tidal response in the atmosphere as well as the sea, but he concluded that it would be too weak to be observed. Given seventeenth century data in Northern Europe, he was certainly correct. The situation is demonstrated in Figure 9.1, which shows time series for surface pressure at both Potsdam (52°N in the present day German Democratic Republic) and Batavia (6°S, the present day capitol of Indonesia, Jakarta). Clearly, whatever tiny tide that might exist at middle latitudes is swamped by large meteorological disturbances[3].

---

[1]Sea breezes, which marginally fit our definition of a tide, were undoubtedly observed earlier.

[2]For readers unfamiliar with sea tides, a simple treatment is provided in Lamb's (1916) classic treatise, *Hydrodynamics*.

[3]Sidney Chapman's (1918) accurate determination of the lunar tide over England was an early triumph of signal detection.

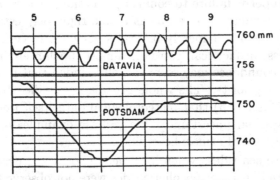

Figure 9.1: Barometric variations (on twofold different scales) at Batavia (6°S) and Potsdam (52°N) during November, 1919. After Bartels (1928).

In the tropics, on the other hand, synoptic scale pressure perturbations are very small, while tidal oscillations are relatively large. The peculiar feature of these atmospheric surface pressure tides is that they are primarily solar semidiurnal. Laplace, already aware of this fact, concluded that the solar dominance implied a thermal origin.

It was Lord Kelvin (1882) who most clearly recognized the paradoxical character of these early observations. First, however, he confirmed the existing data by collecting and harmonically analyzing data from thiry stations for diurnal, semidiurnal, and terdiurnal components. The essence of the paradox is as follows: Gravitational tides are semidiurnal due to the intrinsic semidiurnal character of the forcing; if, however, atmospheric tides are thermally forced, then their forcing is predominantly diurnal. Why then is the response still predominantly semidiurnal? Kelvin put forward the hypothesis that the atmosphere had a free oscillation with zonal *wavenumber* 2 and a period near 12 hours which was resonantly excited by the small semidiurnal component of the thermal forcing. This *resonance hypothesis* dominated thinking on atmospheric tides for almost seventy years. Theoretical work centered on the search for the atmosphere's free oscillations. Following the terminology introduced in Section 8.7, Margules (1890) showed that an atmosphere with an *equivalent depth* of 7.85 km would, indeed, have a free oscillation of the required type. The atmosphere's equivalent depth depends on its thermal structure. In the late nineteenth century, this structure was largely unknown. However, both Rayleigh (1890) and Margules (1890, 1892, 1893), using very crude and unrealistic (in view of today's knowledge) assumptions, concluded that resonance was a possibility.

Lamb (1910, 1916) investigated the matter more systematically. He found that for either an isothermal basic state wherein density variations occur isothermally, or for an atmosphere with a basic state with an adiabatic lapse rate, the equivalent depth was very nearly resonant. Lamb also showed that when the basic state temperature varied linearly (but not adiabatically) with height, the atmosphere had an infinite number of equivalent depths – thus greatly increasing the possibility of resonance. Little note was taken of this result, but another suggestion of Lamb's was followed up: namely, his suggestion that the solar semidiurnal tide might, in fact, be gravitationally forced. His point was that such forcing would require such a degree of resonance to produce the observed tide that it would actually distinguish between the solar semidiurnal period and the lunar semidiurnal period (12 hr 26 min). Lamb, himself, noted at least two problems with this suggestion. First, of course, was the intrinsic unlikelihood of the atmosphere being so highly tuned. The second problem was that the phase of the observed surface pressure tide led rather than lagged the phase of the sun. This was opposite to what calculations showed. Chapman (1924) showed that the last item could be remedied if thermal forcing was of the same magnitude as gravitational forcing. With this rather coarse fix, the resonance theory was largely accepted for the next eight years. In terms of our discussion of scientific methodology, we were still, however, at the stage of simulation rather than theory. This situation changed dramatically with the work of Taylor and Pekeris.

In 1932, G.I. Taylor noted (as we saw in Section 8.7) that an atmosphere with an equivalent depth h would propagate small-scale disturbances (such as would be generated by explosions, earthquakes, etc.) at a speed $\sqrt{gh}$. Using data from the Krakatoa eruption of 1883[4], he showed that the atmospheric pulse travelled at a speed of 319 ms^{-1}, corresponding to $h = 10.4$ km — a value too far from 7.85 km to produce resonance. In 1936, Taylor returned to this problem, having redis-

---

[4]Sometimes old data can serve in place of new data.

covered Lamb's earlier result that the atmosphere might have several equivalent depths. This allowed some hope for the, by now much modified, Kelvin resonance hypothesis. This hope received an immense boost from the work of Pekeris (1937).

Pekeris examined a variety of complicated basic states in order to see what distribution of temperature would support an equivalent depth of 7.85 km[5]. The distribution he found was one where the temperature decreased with height as observed in the troposphere; above the tropopause (*ca.* 12 km) the temperature increased with height to a high value (350°K) near 50 km, and then decreased upwards to a low value. It should be noted that in the mid–1930s we had no direct measurements of upper atmosphere temperature. However, independently of Pekeris, Martyn and Pulley (1936), on the basis of then recent meteor and anomalous sound data, proposed an observationally based thermal structure of the atmosphere which was in remarkable agreement with what Pekeris needed. It was almost as though Pekeris had deduced the atmosphere's complete thermal structure from tidal data at the earth's surface, simply by assuming resonance. His results, moreover, appeared to explain other observations of ionospheric and geomagnetic tidal variations. The vindication of the resonance theory seemed virtually complete. Pekeris countered Taylor's earlier criticism by showing that a low-level disturbance would primarily excite the faster mode associated with $h = 10.4$ km. A reexamination of the Krakatoa evidence by Pekeris even showed some evidence for the existence of the slower mode for which $h = 7.85$ km.

At this point a bit of editorial comment might be in order. If our story were to end at this point, it would have described a truly remarkable scientific achievement. In fact, as we shall soon see, the resonance theory proved to be profoundly wrong. The ability of Pekeris's theory to predict something well beyond the data that had motivated the theory did not end up proving the correctness of the theory! It will be useful to look at the remainder of this story to see where things fell apart. In some ways it is a fairly complicated story. However, before proceding, a few things should be noted concerning Pekeris's work. The explanation of the ionospheric and geomagnetic data did not (on sub-

---

[5]This was an early form of inverse problem.

sequent scrutiny) actually depend on the resonance hypothesis. It is not, however, unheard of in science that one success is used to bolster another unrelated success. Similarly, Pekeris's reexamination of the Krakatoa data demonstrates the very real dangers relating to the analysis of ambiguous and noisy data by theoreticians with vested interests in the outcome of the analysis. Pekeris's claims for the data analysis were modest and circumspect, but even with the best will to be objective, he would have had difficulty not seeing at least hints of what he wanted to see. But all this is jumping ahead of our story. For fifteen years following Pekeris's remarkable work, most research on this subject was devoted to the refinement and interpretation of Pekeris's work. This research is comprehensively reviewed in a monograph by Wilkes (1949). Wilkes's monograph, incidentally, was the standard reference on atmospheric oscillations for over a decade.

The first major objections to the resonance theory emerged in the aftermath of World War II when captured V2 rockets were used to probe the temperature structure of the atmosphere directly. The structure found differed from that proposed by Martyn and Pulley[6]. In particular, the temperature maximum at 50 km was much cooler (about 280°K rather than 350°K). In addition, the temperature decline above 50 km ended around 80 km, above which the temperature again increases, reaching very high values (600–1400°K) above 150 km. Jacchia and Kopal (1951), using an analog computer, investigated the resonance properties of the newly measured temperature profiles. They concluded that with the measured profiles, the atmosphere no longer had a second equivalent depth, and that the magnification of the solar semidiurnal tide was no longer sufficient to account for the observed semidiurnal tide on the basis of any realistic combination of gravitational excitation and excitation due to the upward diffusion of the daily variation of surface temperature[7]. As we shall see at the end of this chapter, Jacchia and Kopal were premature in claiming to have disproven the resonance theory; the real problems had not yet been identified. Nevertheless, their results were widely perceived as constituting the demise

---

[6]Here we see an example of a common phenomenon in meteorology: namely, data that turns out to be not quite data.

[7]Up to this point, this was the only form of thermal forcing considered.

of resonance theory, and this perception fueled the search for additional sources of thermal forcing.

Although most of the sun's radiation is absorbed by the earth's surface, about 10 percent is absorbed directly by the atmosphere, and this appeared a likely source of excitation[8]. Siebert (1961) investigated the effectiveness of insolation absorption by water vapor in the troposphere, and found that it could account for one-third of the observed semidiurnal surface pressure oscillation. This was far more than could be accounted for by gravitational excitation or surface heating. Siebert also investigated the effectiveness of insolation absorption by ozone in the middle atmosphere. He concluded its effect was relatively small. We now know that this last conclusion is wrong. In order to simplify calculations, Siebert used a basic temperature profile which was exceedingly unrealistic above the tropopause. As we shall see later in this chapter, this profile prevented the vertical propagation of semidiurnal tidal oscillations from the stratosphere to the troposphere. Butler and Small (1963) soon corrected this error, and showed that ozone absorption indeed accounted for the remaining two-thirds of the surface semidiurnal oscillation[9].

With a successful and robust theory in hand for the solar semidiurnal tide, we must return to Kelvin's seminal question: Why isn't the diurnal oscillation stronger than the semidiurnal? With the increased data available by the mid–1960s, even this question had become less obvious. Data above the ground up to about 100 km showed that at many levels and latitudes, the diurnal oscillations were as strong and often stronger than semidiurnal oscillations. Lindzen (1967b) carried out theoretical calculations for the diurnal tide which provided sat-

---

[8]The absorptive properties of the atmosphere were actually sufficiently well known in the 1930s. It is curious that no one looked into their possible rôle in generating tides. Undoubtedly, the fact that the atmospheric sciences are actually a small field, and tides a small subset of a small field, played an important part. In addition specialization undeniably encourages a kind of tunnel vision.

[9]In a rough sense, the work of Butler and Small completes our present understanding of the semidiurnal surface oscillation. However, as we shall see, the understanding is by no means complete. There is a discrepancy of about on hour in phase between theory and observation. There is also a problem with the predicted vertical structure of the tidal fields. Recent work suggests that these discrepancies are related to daily variations in rainfall.

isfactory answers for the observed features. Central to the explanation is the fact that on half of the globe (polewards of $\pm 30°$ latitude), 24 hr is longer than the local pendulum day (the period corresponding to the local Coriolis parameter), and under these circumstances a 24 hr oscillation is incapable of propagating vertically. We will explain this behaviour later in this chapter. In any event, because of this, it turns out that 80 percent of the diurnal forcing goes into physically trapped modes which cannot propagate disturbances forced aloft to the ground. The atmospheric response to these modes in the neighbourhood of the excitation is, however, substantial[10]. In addition, there exist (primarily equatorwards of $\pm 30°$ latitude) diurnal modes which propagate vertically. However, as one could deduce from the dispersive properties of internal gravity waves (*viz.* Equation 8.55), the long period and the restricted latitude scale of these waves causes them to have relatively short vertical wavelengths (25 km or less). They are, therefore, subject to some destructive interference effects. Butler and Small suggested, in fact, that this could explain the relatively small amplitude of the diurnal tide, but subsequent calculations showed that this effect would be inadequate. What really proved to be important was that the propagating modes received only 20 percent of the excitation.

The story of tides hardly ends at this point. New data from the upper atmosphere continues to provide challenging questions. Tides still form an interesting focus for both observational and theoretical efforts. Still, after a century, Kelvin's question seems pretty much answered – for the moment.

9.2 Observations

Before proceeding to the mathematical theory of atmospheric tides, it is advisable for us to present a description of the phenomena about which we propose to theorize. As usual, our presentation of the data will be sketchy at best. The data problems discussed in Chapter 5 all apply

---

[10]These trapped diurnal modes were discovered independently by Lindzen (1966) and Kato (1966).

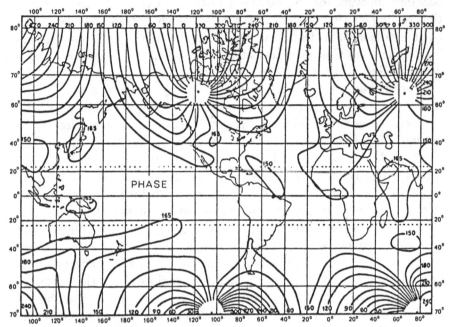

Figure 9.2: World maps showing equilines of phase (σ_2) of $S_2(p)$ relative to local mean time. After Haurwitz (1956).

here as well. Let it suffice to say that at many stages our observational picture is based on inadequate data; in almost all cases, the analyses of data have required the extrication of Fourier components from noisy data, and in some instances even the observational instruments have introduced uncertainties. Details of some of these matters may be found in Chapman and Lindzen (1970).

For many years, almost all data analyses for atmospheric tides were based on surface pressure data. Although tidal oscillations in surface pressure are generally small, at quite a few stations we have as much as 50–100 years of hourly or bi-hourly data. As a result, even today, our best tidal data are for surface pressure. Figures 9.2 and 9.3 show the amplitude and phase of the solar semidiurnal oscillation over the globe; they were prepared by Haurwitz (1956), on the basis of data

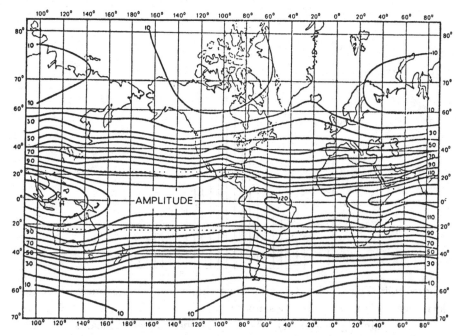

Figure 9.3: World maps showing equilines of amplitude (s_2, unit 10^{-2} mb) of $S_2(p)$. After Haurwitz (1956).

from 296 stations[11]. The phase over most of the globe is relatively constant, implying the dominance of the migrating semidiurnal tide, but other components are found as well (the most significant of which is the semidiurnal standing oscillation for which $s = 0$; *viz.* Figure 9.4). If we let t = local time non-dimensionalized by the solar day, then, according to Haurwitz, the solar semidiurnal tide is well represented by:

$$S_2(p) = \quad 1.16 \ \sin^3 \theta \sin(2\pi t + 158°)$$
$$+ \quad 0.085 \ P_2(\theta) \sin(2\pi t_u + 118°) \ \text{mbar}, \qquad (9.1)$$

---

[11]It is tempting to seek more recent analyses, but since errors decrease only as the square root of the record length, the improvement so far is likely to be pretty negligible.

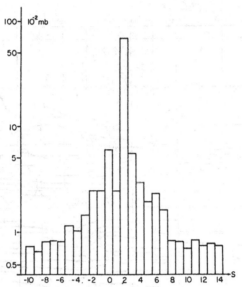

Figure 9.4: The amplitudes (on a logarithmic scale, and averaged over the latitudes 80°N to 70°S) of the semidiurnal pressure waves, parts of $S_2(p)$, of the type $\gamma_s \sin(2t_u + s\phi + \sigma_s)$, where t_u signifies universal mean solar time. After Kertz (1956).

where

$$\theta = \text{colatitude}$$
$$t_u = \text{Greenwich (Universal) time}$$
$$P_2(\theta) = \frac{1}{2}(3\cos^2\theta - 1).$$

One of the remarkable features of $S_2(p)$ is the fact that it hardly varies with season. This can be seen from Figure 9.5. The situation is more difficult for $S_1(p)$. It varies with season, it is weaker, and it is strongly polluted by non-migrating diurnal oscillations (*viz.* Figure 9.6). There are values of s with amplitudes as large as 1/4 of that pertaining to $s = 1$. (For $S_2(p)$, $s = 2$ was twenty times as large as its nearest competitor.) Moreover, large values of s, being associated with a small scale (large gradients), produce larger winds for a given amplitude of pressure oscillation than $s = 1$. We will return to this

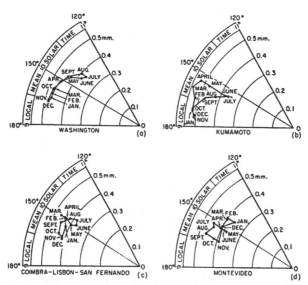

Figure 9.5: Harmonic dials showing the amplitude and phase of $S_2(p)$ for each calendar month for four widely spaced stations in middle latitudes: (a) Washington, D.C.; (b) Kumamoto; (c) mean of Coimbra, Lisbon, and San Fernando; (d) Montevideo (Uruguay). After Chapman (1951).

later. According to Haurwitz (1965), $S_1(p)$ is roughly representable as follows:

$$S_1^1(p) = 593 \, \sin^3 \theta \sin(t + 12°) \, \mu\text{bar}. \tag{9.2}$$

Data have also been analyzed for small terdiurnal and higher harmonics. Even $L_2(p)$ has been isolated. As we can see in Figure 9.7, the amplitude of $L_2(p)$ is about $1/20$ of $S_2(p)$. $L_2(p)$ also has a peculiar seasonal variation, which can only be marginally discerned in Figure 9.8. The seasonal variation occurs with both the Northern and Southern Hemispheres in phase. This is much clearer in Figure 2L.2 in Chapman and Lindzen (1970), which summarizes lunar tidal data from 107 stations.

Data above the surface are rarer and less accurate, but some are available, with relatively recent radar techniques providing useful data well into the thermosphere.

At some radiosonde stations there are sufficiently frequent balloon ascents (four per day) to permit tidal analyses for both diurnal and

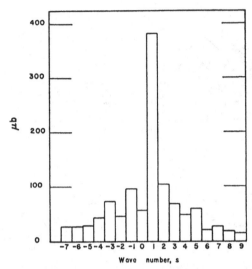

Figure 9.6: The amplitudes (averaged over the latitudes from the North Pole to 60°S) of the diurnal pressure waves, parts of $S_1(p)$, of the type $\gamma_s \sin(t_u + s\phi + \sigma_s)$. After Haurwitz (1965).

semidiurnal components from the ground up to about 10 mb. Early analyses based on such data at a few isolated stations (Harris, Finger, and Teweles, 1962) were unable to distinguish migrating from non-migrating tides, but did establish orders of magnitude. Typically, horizontal wind oscillations were found to have amplitudes $\sim$ 10 cm/sec in the troposphere and $\sim$ 50 cm/sec in the stratosphere. Global pictures of the behaviour of diurnal tides in horizontal wind, based on radiosonde data, were obtained by Wallace and Hartranft (1969) and Wallace and Tadd (1974) using the clever premise that the time average of the difference between wind soundings at 0000 UT and 1200 UT should represent a snapshot of the odd harmonics of the daily variation which are strongly dominated by the diurnal component. Recently, Hsu and Hoskins (1989) have shown that analyzed ECMWF (European Centre for Medium-range Weather Forcasting) data successfully depict diurnal and semidiurnal tides below 50 mb (the upper limit of ECMWF analyses). Their results for the semidiurnal oscillation in zonal wind along the equator are shown in Figure 9.9. We see a clear wavenum-

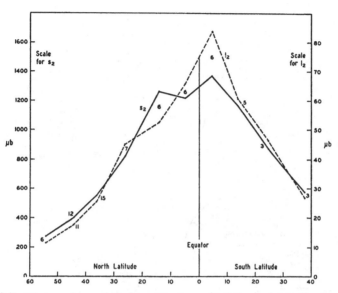

Figure 9.7: Mean values of the amplitude s_2 (full line) and l_2 (broken line) of the annual mean solar and lunar semidiurnal air-tides in barometric pressure, $S_2(p)$ and $L_2(p)$, for $10°$ belts of latitude. The numbers beside each point show from how many stations that point was determined. After Chapman and Westfold (1956).

ber 2 pattern indicative of a migrating tide. We also see very little tilt with height. The diurnal oscillations are more complicated. Figure 9.10 shows a snapshot of the diurnal component of the height field along the equator at 0000 GMT averaged over the winter of 1986/87. There is a clear wavenumber 1 pattern with some evidence of phase tilt. However, the distortion from a strict sine wave has important consequences for the diurnal wind pattern (Why?). This becomes evident in Figure 9.11 which shows a comparably averaged snapshot of the diurnal component of the horizontal wind at 850 mb. The pattern is more complicated; wavenumber 1 is no longer self-evidently dominant. There are numerous regional diurnal circulations. Figure 9.12 shows a similar snapshot at 50 mb; regional features are still evident but less pronounced. As

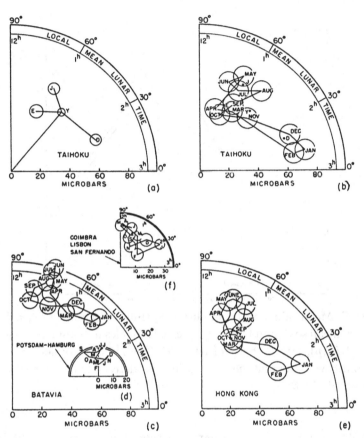

Figure 9.8: Harmonic dials, with probable error circles, indicating the changes of the lunar semidiurnal air-tide in barometric pressure in the course of a year. (a) Annual (y) and four-monthly seasonal (j, e, d) determinations for Taihoku, Formosa (now Taipei, Taiwan) (1897–1932). Also five sets of twelve-monthly mean dial points. After Chapman (1951).

already noted, ECMWF analyses do not extend beyond 50 mb. However, a similar snapshot from Wallace and Hartranft (1969) for 15 mb (involving, however, an annual rather than a winter average), shown in Figure 9.13, does suggest a wavenumber 1 dominance (characterized by flow over the pole; why?). The data up to 15 mb offer some reason to

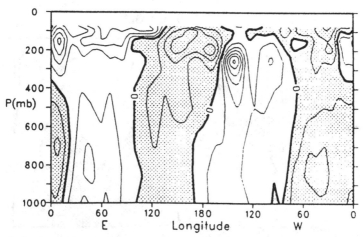

Figure 9.9: Semidiurnal zonal wind field for DJF 1986/87 along the equator; contour interval 0.2 m/sec. Regions of negative values are shaded. After Hsu and Hoskins (1989).

expect that the regional influences die out within the lower stratosphere, and that above 15 mb, diurnal oscillations are mostly migrating.

In the region between 30 and 60 km, most of our data come from meteorological rocket soundings. These are comparatively infrequent and the method of analysis becomes *a priori*, a serious problem. However, it turns out that results of different analyses appear to be compatible (at least for the diurnal component) because tidal winds at these heights are already a very significant part of the total wind (at least in the north–south direction).

This is seen in Figure 9.14, where we show the southerly wind as measured over a period of 51 h at White Sands, N.M. Analyses of tidal waves at various latitudes are now available. Figure 9.15 shows the phase and amplitude of the semidiurnal oscillation at about 30°N. Below 50 km, the results appear quite uncertain (Reed, 1967). In Figures 9.16 and 9.17 we see the diurnal component at 61°N and at 20°N, respectively. Amplitudes are of the order of 10 m sec^{-1} at 60 km but phase at 20°N is more variable than at 61°N (Reed, Oard, and Siemanski, 1969).

0000 GMT

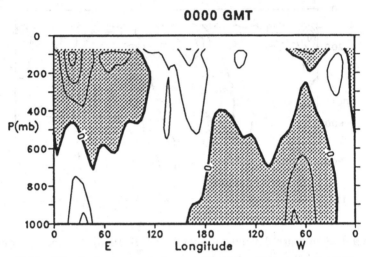

Figure 9.10: Vertical cross section of the diurnal height field along the equator at 0000 GMT in DJF 1986/87. Contour interval is 5 m. Regions of negative values are shaded. After Hsu and Hoskins (1989).

Between 60 km and 80 km, there are too few data for tidal analyses. Between 80 and 105 km, there is a growing body of data from the observation of ionized meteor trails by Doppler radar. The earliest such data were for vertically averaged wind over the whole range 80–105 km. Some such data for Jodrell Bank (Greenhow and Neufeld, 1961) (58 °N) and Adelaide (Elford, 1959) (35 °S) showed typical magnitudes of around 20 m sec^{-1}. All quantities were subject to large seasonal fluctuations and error circles. At Adelaide, diurnal oscillations predominated, whereas at Jordell Bank semidiurnal oscillations predominated; at both stations tidal winds appeared to exceed other winds. The extensive vertical averaging made it difficult to compare these observations with theory. Improvements in meteor radars have made it possible to delineate horizontal winds with vertical resolutions of 1–2 km over the height range 80 to 105 km. Such results are reviewed in Glass and Spizzichino (1974). Typical amplitudes and phases for semidiurnal and diurnal tides obtained by this technique over Garchy, France are shown in Figure 9.18. The semidiurnal phase variation with height is substan-

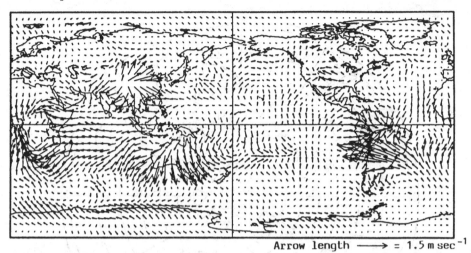

Arrow length ——→ = 1.5 m sec⁻¹

Figure 9.11: The diurnal wind vectors in DJF 1986/87 at 850 mb at 0000 GMT. After Hsu and Hoskins (1989).

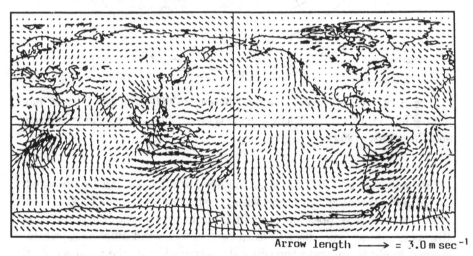

Arrow length ——→ = 3.0 m sec⁻¹

Figure 9.12: The diurnal wind vectors in DJF 1986/87 at 50 mb at 0000 GMT. After Hsu and Hoskins (1989).

tially greater than is typically seen at lower altitudes. The diurnal tide at this location is typically weaker than the semidiurnal tide; it is also usually associated with much shorter vertical wavelengths.

Figure 9.13: Annual average wind differences 0000–1200 GMT at 15 mb plotted in vector form. The length scale is given in the figure. After Wallace and Hartranft, (1969).

Between 90 and 130 km (and higher), wind data can be obtained by visually tracking luminous vapor trails emitted from rockets. In most cases this is possible only in twilight at sunrise and sundown. Hines (1966) used such data to form twelve-hour wind differences, which seemed likely to indicate the diurnal contribution to the total wind at dawn at Wallops Island (38°N). Hines assumed that the average of the winds measured twelve hours apart would be due to the sum of prevailing and semidiurnal winds. His results are shown in Figure 9.19. There is an evident rotation of the diurnal wind vector with height, characteristic of an internal wave with a vertical wavelength of about 20 km. Amplitude appears to grow with height up to 105 km, and then to decay.

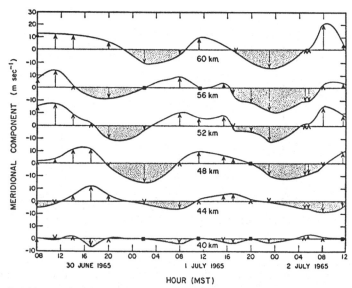

Figure 9.14: Meridional wind component, u, in m/sec averaged over 4 km centred at 40, 44, 48, 52, 56, and 60 km. Positive values indicate a south to north flow. After Beyers, Miers, and Reed (1966).

Over the past twenty years, it has become possible to observe the atmosphere both in the mesosphere and above 100 km in considerable detail using the incoherent backscatter of powerful radar signals. Figure 9.20 shows tidal amplitudes and phases obtained for altitudes between 100 and 450 km over St. Santin, France (45°N). Above 150 km we see that the diurnal component is again dominant; also, above about 225 km all amplitudes and phases are almost independent of height. Finally, it should be noted that the amplitudes are very large ($\sim$ 100 m/s for the diurnal component). The temperature oscillations have comparable amplitudes ($\sim$ 100 K).

Having familiarized ourselves with atmospheric tides as they are actually observed, we will now proceed to their mathematical theory.

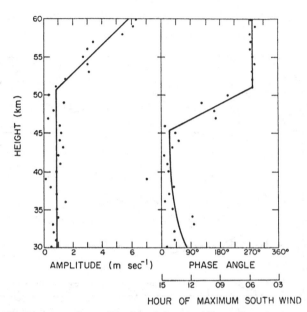

Figure 9.15: Phase and amplitude of the semidiurnal variation of the meridional wind component u at 30°N based on data from White Sands (32.4°N) and Cape Kennedy (28.5°N). After Reed (1967).

9.3   Theory

We will restrict ourselves to 'migrating tides' whose dependence on time and longitude is given by $e^{2\pi i s t_\ell}$ where t_ℓ = local time in days. $t_\ell = t_u + \phi/2\pi$, where ϕ is longitude and t_u = universal time. We will generally refer to t_u simply as t. $s = 1$ corresponds to a diurnal tide; $s = 2$ refers to a semidiurnal tide, and so forth. Such oscillations have phase speeds equal to the linear rotation speed of the earth. Since this speed is generally much larger than typical flow speeds we usually assume the basic state to be static. Also the periods are sufficiently long to allow us to use the hydrostatic approximation. This, in turn, allows us to replace z as a vertical coordinate with

$$z^* \equiv -\ln\left(\frac{p}{p_s}\right). \tag{9.3}$$

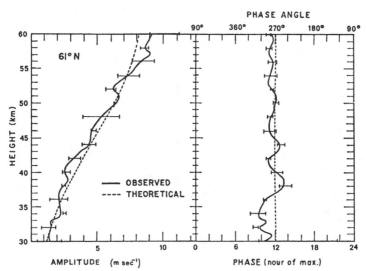

Figure 9.16: Phase and amplitude of the diurnal variation of the meridional wind component u at 61°N. Phase angle, in accordance with the usual convention, gives the degrees in advance of the origin (chosen as midnight) at which the sine curve crosses from − to +. The theoretical curves will be discussed later in this chapter. After Reed *et al.* (1969).

(Recall $p = p_s e^{-z^*}$, where $z^* = \int_0^z \frac{dz}{H}, H = \frac{RT_0}{g}$). This coordinate system (log–pressure) is described in Holton. The resulting equations no longer formally include density, and, as a result, they are virtually identical to the Boussinesq equations without, however, the same restrictions. In this coordinate system vertical velocity is replaced by

$$w^* = \frac{dz^*}{dt} = -\frac{1}{p}\frac{dp}{dt} \qquad (9.4)$$

and pressure is replaced by geopotential

$$\Phi = gz(z^*).$$

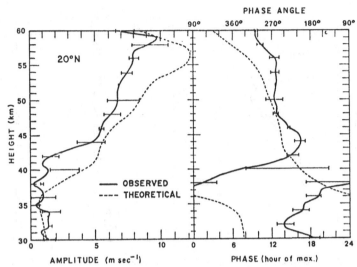

Figure 9.17: Phase and amplitude of the diurnal variation of the meridional wind component u at 20°N. Phase angle, in accordance with the usual
convention, gives the degrees in advance of the origin (chosen as midnight)
at which the sine curve crosses from − to +. The theoretical curves will be
discussed later in this chapter. After Reed *et al.* (1969).

The only (minor) difficulty with this scheme is that the lower boundary
condition is that

$$w = 0 \quad \text{at } z = 0 \tag{9.5}$$

and w^* is *not* the vertical velocity.

The correct lower boundary condition in $\log -p$ coordinates is obtained as follows:

At $z = z^* = 0$, $w = 0$, so that

$$w^* = -\frac{1}{p_s}\frac{dp'}{dt} = -\frac{1}{p_s}\left(\frac{\partial p'}{\partial t} + w'\frac{dp_0}{dz}\right) = -\frac{1}{p_s}\frac{\partial p'}{\partial t}.$$

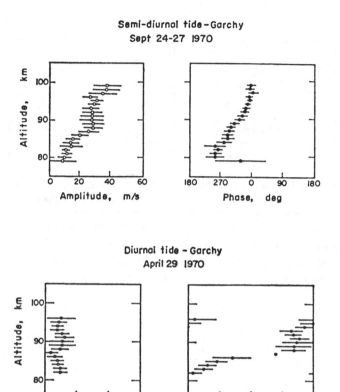

Figure 9.18: Amplitude and phase of the semidiurnal component of the eastward velocity over Garchy observed by meteor radar during September 24–7, 1970 (top), and of the diurnal component during April 29, 1970 (bottom). After Glass and Spizzichino (1974).

From $dp = \frac{\partial p}{\partial t}dt + \cdots + \frac{\partial p}{\partial z}dz$ we get

$$\left(\frac{\partial z}{\partial t}\right)_p = -\frac{\frac{\partial p'}{\partial t}}{\frac{\partial p}{\partial z}} = \frac{\frac{\partial p'}{\partial t}}{\rho g}$$

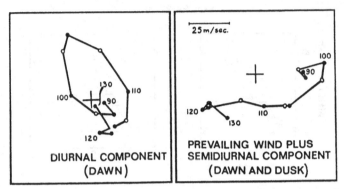

Figure 9.19: Vector diagrams showing (a) the diurnal tide at dawn and (b) the prevailing wind plus the semidiurnal tide at its dawn-dusk phase, as functions of height. The data used were from both Wallops Island, Virginia, and from Sardinia (both near 38°N). After Hines (1966).

or

$$\frac{\partial p'}{\partial t} = \rho \frac{\partial \Phi}{\partial t}.$$

Thus

$$w^* = -\frac{\rho_s}{p_s}\frac{\partial \Phi}{\partial t} = -\frac{1}{gH(0)}\frac{\partial \Phi'}{\partial t} \text{ at } z^* = 0. \tag{9.6}$$

Equation 9.6 is our appropriate lower boundary condition.

It is, unfortunately, the case that tidal theory has used different horizontal coordinates than those used in the rest of meteorology; $\theta =$ colatitude, $\phi =$ longitude, $u =$ northerly velocity, and $v =$ westerly velocity. Assuming time and longitude dependence of the form $e^{i(\sigma t + s\phi)}$ (somewhat more general than our earlier choice) our linearized equations for horizontal motion are simply

$$i\sigma u' - 2\Omega\cos\Theta v' = -\frac{1}{a}\frac{\partial}{\partial\theta}\Phi' \tag{9.7}$$

and

$$i\sigma v' + 2\Omega\cos\theta u' = -\frac{is}{a\sin\theta}\Phi'. \tag{9.8}$$

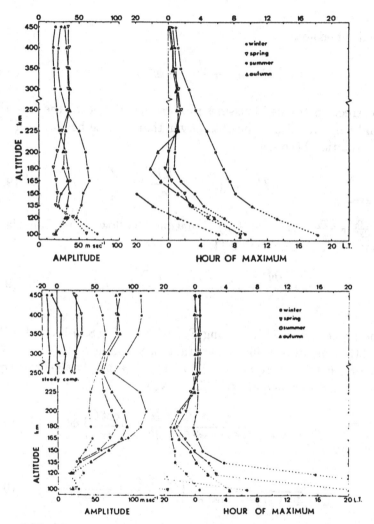

Figure 9.20: Mean seasonal vertical structures of amplitude and phase of the southward neutral wind from 1971–2 observations at St. Santin (45°N). (*top*) Semidiurnal component; (*bottom*) Steady and diurnal components. After Amayenc (1974).

The hydrostatic relation becomes

$$\frac{\partial \Phi'}{\partial z^*} = RT'. \tag{9.9}$$

Continuity becomes

$$\nabla \cdot \vec{u}_{hor} + \frac{\partial w^*}{\partial z^*} - w^* = 0 \tag{9.10}$$

(The correction to the Boussinesq expression is due to the fact that our fluid can extend over heights larger than a scale height.), and the energy equation becomes

$$i\sigma T' + w^*\left(\frac{dT_0}{dz^*} + \frac{RT_0}{c_p}\right) = \frac{J}{c_p}. \tag{9.11}$$

(N.B. $\frac{dT_0}{dz^*} + \frac{RT_0}{c_p} = H(\frac{dT_0}{dz} + \frac{g}{c_p})$.) Equation 9.9 allows us to immediately eliminate T' from Equation 9.11:

$$i\sigma\frac{\partial \Phi'}{\partial z^*} + w^*R\left(\frac{dT_0}{dz^*} + \frac{RT_0}{c_p}\right) = \kappa J. \tag{9.12}$$

The procedure used in solving Equations 9.7, 9.8, 9.10, and 9.12 is sufficiently general in utility to warrant sketching here.

We first note that Equations 9.7 and 9.8 are simply algebraic equations in u' and v' which are trivially solved:

$$u' = \frac{i\sigma}{4a^2\Omega^2(f^2 - \cos^2\theta)}\left(\frac{\partial}{\partial\theta} + \frac{s\cot\theta}{f}\right)\Phi' \tag{9.13}$$

and

$$v' = \frac{-\sigma}{4a^2\Omega^2(f^2 - \cos^2\theta)}\left(\frac{\cos\theta}{f}\frac{\partial}{\partial\theta} + \frac{s}{\sin\theta}\right)\Phi', \tag{9.14}$$

where

$$f \equiv \sigma/2\Omega.$$

Now u' and v' (and all information about rotation and sphericity) enter the remaining two equations only through $\nabla \cdot \vec{u}_{hor}$ in Equation 9.10.

Using Equations 9.13 and 9.14 we may express $\nabla \cdot \vec{u}_{hor}$ as follows:

$$\nabla \cdot \vec{u}_{hor} \quad = \frac{1}{a \sin \theta} \frac{\partial}{\partial \theta}(u' \sin \theta) + \frac{1}{a \sin \theta} isv' \qquad (9.15)$$

$$= \frac{i\sigma}{4a^2 \Omega^2} F[\Phi'], \qquad (9.16)$$

where

$$F\,[\Phi'] \quad \equiv \left\{ \frac{1}{\sin \theta} \frac{\partial}{\partial \theta} \left(\frac{\sin \theta}{f^2 - \cos^2 \theta} \frac{\partial}{\partial \theta} \right) \right.$$

$$\left. - \frac{1}{f^2 - \cos^2 \theta} \left(\frac{s}{f} \frac{f^2 + \cos^2 \theta}{f^2 - \cos^2 \theta} + \frac{s^2}{\sin^2 \theta} \right) \right\} \Phi'. \qquad (9.17)$$

9.3.1 Laplace's tidal equation

Note that apart from F, Equations 9.10 and 9.12 depend only on z^*. We can make the present problem almost identical to the problem in Section 8.7 by separating variables so that

$$\nabla \cdot \vec{u}_{hor} = -\frac{i\sigma}{gh} \Phi'$$

or more correctly

$$\frac{i\sigma}{4a^2 \Omega^2} F[\Theta_n] = -\frac{i\sigma}{gh_n} \Theta_n. \qquad (9.18)$$

For each σ and s we have an infinitude of north-south modes – each with an equivalent depth, h_n, exactly analogous to our earlier example. Equation 9.18 is *Laplace's tidal equation*. It defines an eigenfunction-eigenvalue problem where the equivalent depths are the eigenvalues and the eigenfunctions are known as Hough functions. Hough functions play a major role in meteorology and oceanography – representing as they do very general classes of oscillations including gravity waves, Rossby waves, and mixtures.

Solving Equation 9.18 is a technical task which we will skip over. (Details may be found in Chapman and Lindzen, 1970) We will, instead, look at the counterpart of Equation 9.18 for simpler geometries, in order that we may understand results obtained with Equation 9.18. For the moment we should note that all information about geometry and rotation is contained in h_n.

9.3.2 Vertical structure equation

Formally, the equation for z^* dependence will be the same regardless of geometry. Substituting (9.18) into (9.10) and (9.12) (and expanding J, w^*, and Φ' in terms of $\Theta_n(\theta)$) we get

$$-\frac{i\sigma}{gh_n}\Phi'_n + \frac{dw_n^*}{dz^*} - w_n^* = 0 \qquad (9.19)$$

$$i\sigma\frac{d\Phi'_n}{dz^*} + w_n^* R\left(\frac{dT_0}{dz^*} + \frac{RT_0}{c_p}\right) = \kappa J_n, \qquad (9.20)$$

from which Φ'_n is readily eliminated to give

$$\frac{d^2 w_n^*}{dz^{*2}} - \frac{dw_n^*}{dz^*} + w_n^*\frac{R}{gh_n}\left(\frac{dT_0}{dz^*} + \frac{RT_0}{c_p}\right) = \frac{\kappa J_n}{gh_n}. \qquad (9.21)$$

If we let

$$w^* = \tilde{w}e^{z^*/2}, \qquad (9.22)$$

(9.21) becomes

$$\frac{d^2\tilde{w}_n}{dz^{*2}} + \left\{\underbrace{\frac{R}{gh_n}\left(\frac{dT_0}{dz^*} + \frac{RT_0}{c_p}\right)}_{\frac{1}{h_n}(\frac{dH}{dz^*}+\kappa H)} - \frac{1}{4}\right\}\tilde{w}_n = \frac{\kappa J_n}{gh_n}e^{-z^*/2}. \qquad (9.23)$$

Using (9.19), our l.b.c. (i.e., lower boundary condition), (9.6) becomes

$$\frac{dw_n^*}{dz^*} + \left(\frac{H(0)}{h_n} - 1\right)w_n^* = 0 \text{ at } z^* = 0, \qquad (9.24)$$

or

$$\frac{d\tilde{w}_n}{dz^*} + \left(\frac{H}{h_n} - \frac{1}{2}\right)\tilde{w}_n = 0 \text{ at } z^* = 0. \qquad (9.25)$$

Again our upper boundary condition is a radiation condition. Note the following:

(a) From (9.22) we see that vertically propagating waves increase in amplitude with height in such manner as to leave energy density constant.

(b) The higher a given thermal forcing ($\rho J \sim$ constant) is applied the greater the response every place. (You have an exercise on this.)

(c) The equivalent depth of the atmosphere is the eigenvalue of Equations 9.25, 9.23 (with $J = 0$), and the upper boundary condition. As an exercise you will show that when $T_0 = $ constant, there is only one atmospheric equivalent depth, $h = \gamma H$.

(d) The equivalent depth of a mode determines the vertical wavenumber.

9.3.3 Simplified Laplace's tidal equation

Let us now look at the counterpart of Laplace's tidal equation on a rotating planar channel. Our equations for horizontal motion are (assuming solutions of the form $e^{i(\sigma t + kx)}$)

$$i\sigma u' - fv' = -ik\Phi' \tag{9.26}$$

$$i\sigma v' + fu' = -\frac{\partial \Phi'}{\partial y}, \tag{9.27}$$

where

$$v' = 0 \text{ at } y = 0, L.$$

From (9.26) and (9.27)

$$u' = \frac{\sigma k\Phi' - f\frac{\partial \Phi'}{\partial y}}{(f^2 - \sigma^2)} \tag{9.28}$$

$$v' = \frac{ikf\Phi' - i\sigma\frac{\partial \Phi'}{\partial y}}{(f^2 - \sigma^2)} \tag{9.29}$$

$$\nabla \cdot \vec{u}'_{hor} = \frac{\partial u'}{\partial x} + \frac{\partial v'}{\partial y} = \frac{-i\sigma}{(f^2 - \sigma^2)}\left\{ \frac{\partial^2 \Phi'}{\partial y^2} - k^2\Phi' \right\} \tag{9.30}$$

'Laplace's tidal equation' becomes

$$\frac{-i\sigma}{f^2 - \sigma^2}\left\{\frac{\partial^2 \Theta_n}{\partial y^2} - k^2\Theta_n\right\} = -\frac{i\sigma}{gh_n}\Theta_n$$

or

$$\frac{d^2\Theta_n}{dy^2} - \left\{\frac{f^2 - \sigma^2}{gh_n} + k^2\right\}\Theta_n = 0, \tag{9.31}$$

where

$$\frac{d\Theta_n}{dy} - \frac{kf}{\sigma}\Theta_n = 0 \text{ at } y = 0, L. \tag{9.32}$$

If we write

$$\Theta_n = \sin \ell y + A\cos \ell y$$

then (9.32) becomes

$$\ell\cos \ell y - A\ell\sin \ell y - \frac{kf}{\sigma}\sin \ell y - A\frac{kf}{\sigma}\cos \ell y = 0, \ y = 0, L,$$

which in turn implies

$$A = \frac{\sigma\ell}{kf}$$

and

$$\sin \ell y = 0 \text{ at } y = L,$$

or

$$\ell_n = \frac{n\pi}{L}.$$

Equation 9.31 now gives us an expression for h_n:

$$-\ell_n^2 - \left\{\frac{f^2 - \sigma^2}{gh_n} + k^2\right\} = 0,$$

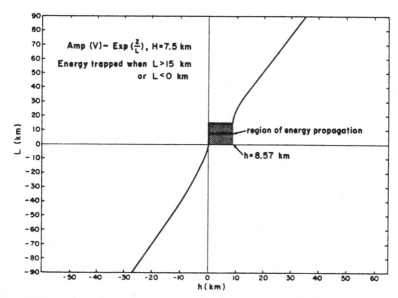

Figure 9.21: Energy trapping as a function of equivalent depth (see text for details). After Lindzen (1967b).

or

$$gh_n = \frac{\sigma^2 - f^2}{k^2 + (\frac{n\pi}{L})^2}. \tag{9.33}$$

Note the following:

(a) If we set $f = 0$ and $n = 0$, we recover our earlier results for internal gravity waves (restricted by hydrostaticity – but extended to a deep fluid).

(b) h_n is positive only if $\sigma^2 > f^2$; if $\sigma^2 < f^2$, h_n is negative.

(c) From Equation 9.23 (known as the vertical structure equation) we see that negative h_n is associated with vertical trapping. What this means, physically, is that at long periods, geostrophic balances are established faster than the oscillatory cross isobaric response.

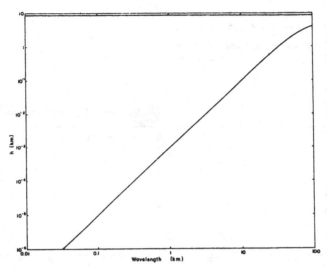

Figure 9.22: Vertical wavelength as a function of equivalent depth. After Lindzen (1967b).

Recalling our earlier discussion of the vertical structure equation, note that h_n determines the vertical wavelength of a given mode

$$m^2 \approx \frac{1}{h_n H}\left(\frac{1}{H}\frac{dH}{dz^*} + \kappa\right) - \frac{1}{4H^2} \tag{9.34}$$

$$\approx \frac{\kappa}{h_n H} - \frac{1}{4H^2} \text{ for an isothermal basic state,}$$

$$VWL \equiv \frac{2\pi}{m}.$$

Figures 9.21 and 9.22 from Lindzen (1967b) show the relation between h_n and VWL (vertical wavelength). From (9.33) we see that h_n (and hence VWL) decreases as n increases (and/or L decreases).

9.3.4 Overall procedure

Finally, we must return to tides. The following flow chart reviews our procedure.

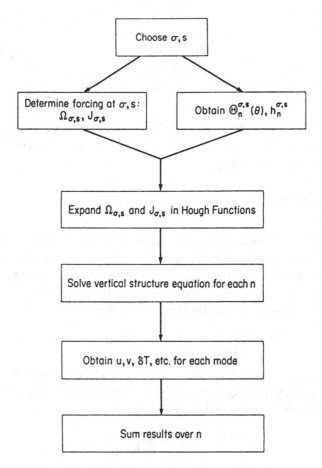

The questions we wish to focus on are:

1. Why is the semidiurnal surface pressure oscillation stronger and more regular than the diurnal oscillation?

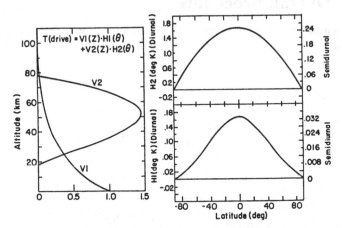

Figure 9.23: Vertical distributions of thermal excitation due to water vapor
(V1) and ozone (V2); latitude distributions for water vapor (H1) and ozone
(H2). After Lindzen (1968).

2. Can we account for the specific observed magnitudes and structures?

In addition, we will take a brief look at the lunar tides – not because
they are important *per se*, but because they tell us something very
significant about how the atmosphere responds to forcing. We note
here, for reference purposes, that the only effect of gravitational forcing
on our equations is to modify the lower boundary condition

$$\frac{d\tilde{w}_n}{dz^*} + \left(\frac{H}{h_n} - \frac{1}{2}\right)\tilde{w}_n = \frac{i\sigma}{gh_n}\Omega_n \quad \text{at } z^* = 0, \tag{9.35}$$

where Ω_n is a tidal contribution to the gravitational potential.

The thermal forcing for diurnal and semidiurnal tides is shown in
Figure 9.23. It is expressed in terms of

$$T = \frac{\kappa J}{i\sigma R}. \tag{9.36}$$

This is the temperature amplitude that would be produced by J in the
absence of dynamics. For the diurnal component, T is maximum at
1800 LT, while for the semidiurnal component T has maxima at 0300
and 1500 LT.

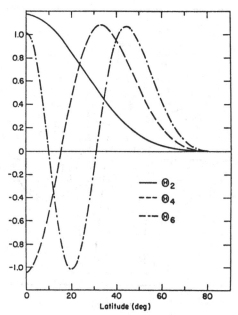

Figure 9.24: Latitude distribution for the first three symmetric solar semidiurnal migrating Hough functions. After Chapman and Lindzen (1970).

9.3.5 Semidiurnal and diurnal solutions – Hough functions

The Hough functions for the semidiurnal tide are shown in Figure 9.24; those for the diurnal tide are shown in Figure 9.25; the equivalent depths are shown in Table 9.1. Notice that the Hough functions for S_2 smoothly span the globe. The main mode resembles the latitude structure of the forcing and has an equivalent depth, 7.85 km, which is associated with either an almost infinite VWL or, sometimes, mild trapping. If we expand the heating functions we get

$$H_{O_3}^{S-D} = 0.25°K\Theta_2^{S-D} + 0.065°K\Theta_4^{S-D} + 0.036°K\Theta_6^{S-D} + \ldots \quad (9.37)$$

$$H_{H_2O}^{S-D} = 0.031°K\Theta_2^{S-D} + 0.008°K\Theta_4^{S-D} + 0.0045°K\Theta_6^{S-D} + \ldots, \quad (9.38)$$

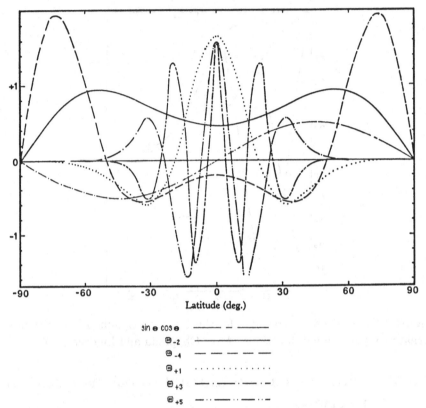

sin θ cos θ	— — —
Θ_{-2}	———————
Θ_{-4}	— — — — —
Θ_{+1}	··············
Θ_{+3}	—·—·—·—··
Θ_{+5}	—··—··—··—

Figure 9.25: Symmetric Hough functions for the migrating solar diurnal thermal tide. Also shown is $\sin\theta\cos\theta$, the most important odd mode. After Lindzen (1967b).

that is, the forcing primarily excites Θ_2^{S-D}. The situation for S_1 is very different indeed. Here we see two distinct sets of eigenfunctions: one concentrated in latitudes poleward of 30° latitude with negative equivalent depths and one concentrated equatorward of 30° with small positive equivalent depths. Our previous discussion enables us to understand these results. From (9.33) we see that negative (positive) equivalent depths are associated with $\sigma^2 < f^2(\sigma^2 > f^2)$.

Table 9.1:

Diurnal Mode #	h_n	Semidiurnal Mode #	h_n
+ 1	.69 km	2	7.85 km
+ 3	.12 km	4	2.11 km
+ 5	.05 km	6	0.96 km
- 2	-12.27 km		
- 4	-1.76 km		

On a sphere we have

$$f = 2\Omega \sin \phi$$

and for S_1, $\sigma = \Omega$. Therefore

$$\sigma = f$$

when

$$\sin \phi = \frac{1}{2},$$

or

$$\phi = 30°.$$

Global modes are formed such that poleward of 30° negative equivalent depth modes oscillate meridionally (real ℓ) and equatorward of 30° they decay exponentially (imaginary ℓ). The opposite situation obtains for positive equivalent depth modes. In addition, the positive (propagating) modes, being confined to the region $\mid \phi \mid < 30°$, have much larger effective ℓs (meridional wavenumbers) and smaller hs than would global modes. This matter will be elucidated in an exercise.

The expansion of diurnal heating yields

$$H_{0_3}^D = \quad 1.63°K\Theta_{-2}^D - 0.51°K\Theta_{-4}^D + \ldots$$

$$+ \quad 0.54°K\Theta_1^D - 0.14°K\Theta_3^D + \ldots \tag{9.39}$$

and

$$H^D_{\text{H}_2\text{O}} = \quad 0.16°\text{K}\Theta^D_{-2} - 0.055°\text{K}\Theta^D_{-4} + \ldots$$

$$+ \quad 0.062°\text{K}\Theta^D_1 - 0.016°\text{K}\Theta^D_3 + \ldots \qquad (9.40)$$

The above results provide an immediate answer to our first question: $S_1(p_s)$ is weaker because most of the forcing goes into trapped modes which do not effectively influence the ground. It is irregular largely because the surface response involves higher order modes which are more susceptible to regional variations. (Note that winds associated with higher order modes may even be larger.) By contrast, $S_2(p_2)$ receives almost all its forcing in a single global mode which is insensitive to regional fluctuations. Moreover, the main S-D mode has an almost infinite VWL so that all forcing contributes 'in phase'[12].

With respect to the second question we obtain $S_2(p_s) \sim 1.1$ mb (with two-thirds of this coming from O_3) with maxima at 0900 and 2100 LT. The amplitude is about right but the observed maxima occur at 0940 and 2140 LT. For $S_1(p_s)$, it is more useful to look at the Hough decomposition

$$S_1^{\text{H}_2\text{O}}(p_s) \quad = \quad \{137\Theta^D_{-2} - 68\Theta^D_{-4} + \ldots$$

$$+ \quad 117e^{56°i}\Theta^D_1 - 13e^{73°i}\Theta^D_3 + \ldots\}e^{i(\Omega t + \phi)}\mu\text{b} \quad (9.41)$$

$$S_1^{O_3}(p_s) \quad = \quad \{44\Theta^D_{-2} - 3.4\Theta^D_{-4} + \ldots$$

$$+ \quad 94e^{13°i}\Theta^D_1 - 3.75e^{16°i}\Theta^D_3 + \ldots\}e^{i(\Omega t + \phi)}\mu\text{b} \quad (9.42)$$

---

[12]On page 166 we remarked that Siebert (1961) had chosen a temperature profile which suppressed the propagation of the semidiurnal wave excited by ozone heating. What Siebert did was to choose a distribution of T_0 such that $\frac{dH}{dz^*} + \kappa H =$ constant (*viz.* Equation 9.23). Such a T_0 decreases with height in the troposphere – reasonably enough. However, the profile approaches asymptotically to a very cold constant temperature above the troposphere. This cold temperature leads to m^2 being significantly negative – as opposed to being almost zero (*viz.* Equation 9.34). Thus ozone forcing is prevented from affecting the surface pressure. Butler and Small (1963) used a realistic profile for T_0 which does not have this problem.

The sum of Equations 9.41 and 9.42 reasonably accounts for observations. Note the relative suppression of trapped modes. Overall, the largest contributor to $S_1(p_s)$ is the ineffectively excited first propagating mode. Note also that at least three modes are of comparable importance.

We will not go into a detailed discussion of the theoretical results for upper air fields, but Figures 9.16 and 9.17 show remarkable agreement between theory and observation. Note that phase variation with height, which is evident at 20° latitude, is virtually absent at 60° latitude (Why?). Figure 9.26 shows theoretical results for semidiurnal northerly velocity oscillations. A comparison with Figure 9.15 shows compatibility with observed magnitudes but the theory predicts a 180° phase shift near 28 km while it is observed at much greater heights. Interestingly, both this discrepancy and that in the phase of $S_2(p_s)$ led to the recognition that an additional important source of tidal forcing arises from the daily variations in tropical rainfall (Lindzen, 1978).

9.3.6 Lunar semidiurnal tide

Finally, we turn briefly to the lunar tide $L_2(p_s)$. Its Hough functions are much like those for S_2. The equivalent depth of its main mode is 7.07 km. Recall that for an isothermal basic state the atmosphere has a single equivalent depth, $h = \gamma H \approx 11$ km, which is far from resonance for the main semidiurnal modes. However, for $h \sim 7$ km we see from Equation 9.23, the vertical structure equation, that the local vertical wavenumber hovers around zero, and varies with height; there is a turning point near 60 km. Thus, as we noted, additional equivalent depths might exist and resonance might be possible (remember the behaviour of a fluid with a lid). This possibility has, in fact, been dismissed too casually. Theoretically, one finds that one can predict the observed $L_2(p_s)$ with an isothermal basic state, but Sawada (1956) found that for different basic T_0s, shown in Figure 9.27, responses shown in Figure 9.28 were obtained. Such extreme variability is certainly characteristic of resonance. Now two points must be made

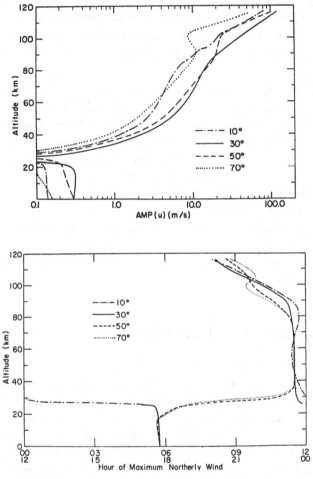

Figure 9.26: (*Top*) Amplitude of the solar semidiurnal component of u at various latitudes: equatorial standard atmosphere is used for $T_0(z)$. (*Bottom*) Phase (hour of maximum) of the solar semidiurnal components of u at various latitudes. After Lindzen (1968).

(a) No such sensitivity is found for the thermally forced tides; and

(b) No such extreme variability is observed for $L_2(p_s)$. So what is happening? First, resonance of an internal wave requires that

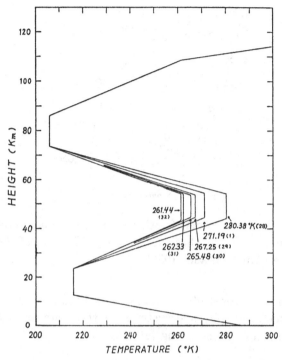

Figure 9.27: Various temperature profiles used in calculating the lunar semidiurnal surface pressure oscillation. The maximum temperature of the statopause and a profile number are shown for each of the profiles. After Sawada (1956).

a wave travel up and down at least several times between the ground and a turning point in such a manner as to produce coherent interference. This is possible for forcing at a single level – but not for a distributed thermal excitation. (Think back to the resonance exercises at the end of Chapter 8.) Even, however, with forcing at the ground, the surface constituting the turning point must be horizontal. In reality, the basic temperature varies with latitude and coherent reflections are difficult to achieve. This

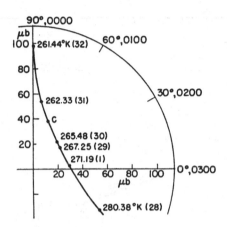

Figure 9.28: A harmonic dial for the lunar semidiurnal surface pressure oscillation. Amplitude and phase are shown as functions of the basic temperature profile. After Sawada (1956).

shows, rather generally, the very unlikely nature of internal wave resonance in geophysical systems. A thorough analysis of this is given in Lindzen and Hong (1974).

Exercises

9.1 Show that Equation 9.23 with $J = 0$ and $T_0 =$ constant has a solution which satisfies Equation 9.25 when $h = \gamma H$. This solution is called a 'Lamb wave'. For a planar geometry where $f = 0$ and $v' = 0$ (solution independent of y; *viz.* Equations 9.26 and 9.27) show that the Lamb wave is a horizontally propagating acoustic wave. Discuss how the equations in $\log -p$ coordinates, which formally look like the equations for an incompressible fluid, can still have acoustic waves. Hint: The answer lies in the lower boundary condition.

9.2 Consider a thermally forced oscillation (*viz.* Equation 9.23) where the forcing consists of a given energy oscillation over a thin layer of thickness Δz; that is,

$$\rho J \Delta z = Q = \text{constant}.$$

For simplicity, let $T_0 = $ constant and let $\Delta z \ll VWL$. If the forcing is centred at a height z_f, show that the response everywhere increases exponentially with z_f. How is this possible?

9.3 In Equation 9.33, $\sigma = \Omega$ for the diurnal tide and $\sigma = 2\Omega$ for the semidiurnal tide. At the equator $f = 0$ and $k = s/a$ ($a = $ radius of Earth), where $s = 1$ for the diurnal tide and $s = 2$ for the semidiurnal tide. For the positive equivalent depths in Table 10.1 calculate ℓ_n from Equation 9.33 and compare your results with the Hough functions shown in Figures 9.24 and 9.25. For the negative equivalent depth modes take $f = 2\Omega \sin 60°$ and $k = 1/a \cos \phi = 1/a \cos 60°$ and calculate the ℓ_ns for the negative h_ns given in Table 9.1. Again, compare your results with actual Hough functions. (N.B. These are local approximations, so you should compare your results in the neighbourhood of $0°$ (or $60°$) with the Hough functions in the same neighbourhood.)

Chapter 10

Internal gravity waves, 2

Basic states with shear

Supplemental reading:

Lindzen (1981)

Lindzen and Holton (1968)

Plumb (1977)

10.1 WKBJ analysis

Most waves in the atmosphere and ocean have phase speeds of the same order as the basic flow. In contrast to tides, variations in the basic flow will be of major importance to such waves. Moreover, waves can strongly affect mean flows. Again, we will choose to study this situation in the simplest relevant configuration. We will ignore rotation and retain $\log -p$ coordinates. Our basic flow will depend only on z^*, and we will consider perturbations for which $v' = 0$. Our equations are

$$\frac{\partial u'}{\partial t} + U_0(z^*)\frac{\partial u'}{\partial x} + w^*\frac{dU_0}{dz^*} + \frac{\partial \Phi'}{\partial x} = -au' \tag{10.1}$$

$$\frac{\partial u'}{\partial x} + \frac{\partial w^*}{\partial z^*} - w^* = 0 \tag{10.2}$$

$$\frac{\partial \Phi'}{\partial z^*} = RT' \tag{10.3}$$

$$\frac{\partial T'}{\partial t} + U_0\frac{\partial T'}{\partial x} + w^*\left(\frac{dT_0}{dz^*} + \kappa T_0\right) = -aT', \tag{10.4}$$

where a = linear damping rate. If we look for solutions of the form $e^{ik(x-ct)}$, 'a' can simply be taken as a modification of 'c', so for the moment we will simply forget the damping. Assuming harmonic dependence on x and t, Equations 10.1–10.4 become

$$-ik(c - U_0)u' + w^*\frac{dU_0}{dz^*} + ik\Phi' = 0 \tag{10.5}$$

$$iku' + \frac{dw^*}{dz^*} - w^* = 0 \tag{10.6}$$

$$-ik(c - U_0)\frac{d\Phi'}{dz^*} + w^*R\left(\frac{dT_0}{dz^*} + \kappa T_0\right) = 0. \tag{10.7}$$

Eliminating u' and Φ' yields

$$(c - U_0)\left\{(c - U_0)\left(\frac{d^2w^*}{dz^{*2}} - \frac{dw^*}{dz^*}\right) + w^*\frac{dU_0}{dz^*} + w^*\frac{d^2U_0}{dz^{*2}}\right\}$$

$$+w^*R\left(\frac{dT_0}{dz^*} + \kappa T_0\right) = 0.$$

Introducing $\tilde{w} = w^*e^{-z^*/2}$ again and organizing terms we get

$$\frac{d^2\tilde{w}}{dz^{*2}} + \left\{\underbrace{\frac{R(\frac{dT_0}{dz^*} + \kappa T_0)}{(c - U_0)^2}}_{A} + \underbrace{\frac{(\frac{d^2U_0}{dz^{*2}} + \frac{dU_0}{dz^*})}{(c - U_0)}}_{B} - \frac{1}{4}\right\}\tilde{w} = 0. \tag{10.8}$$

Only term B is new; term A differs from earlier results only in that a constant c has been replaced by a variable $c - U_0(z^*)$. If $U_0(z^*)$ is varying 'slowly' then we may ignore B (though it will play a crucial rôle in certain instability problems). Let

$$\lambda^2 \equiv \left\{ \frac{R(\frac{dT_0}{dz^*} + \kappa T_0)}{(c - U_0)^2} - \frac{1}{4} \right\}. \tag{10.9}$$

When λ^2 is constant the solution of (10.8) is trivial. When it is varying slowly we have an asymptotically approximate solution in the form of the WKBJ approximation which is almost as trivial[1]:

$$w^* \approx \frac{A e^{z^*/2}}{\lambda^{1/2}} \exp\left\{ -i \int^{z^*} \lambda dz^* \right\}. \tag{10.10}$$

The replacement of $i\lambda z^*$ by $i \int^{z^*} \lambda dz^*$ is intuitively obvious. The slowly varying amplitude factor, $\lambda^{-1/2}$ will be discussed later. Equation 10.10 gives us a basis for assessing the effects of the basic state on a wave. We have already noticed the profound effect of density stratification in producing exponential growth with height. *This almost certainly accounts for the increasing prominence of internal gravity waves in the upper atmosphere.* For the moment, however, we wish to concentrate on the changes produced by variations in U_0. If we concentrate on large values of λ^2,

$$\lambda^2 \approx \frac{R(\frac{dT_0}{dz^*} + \kappa T_0)}{(c - U_0)^2} = \frac{H^2 N^2}{(c - U_0)^2} \tag{10.11}$$

$$\lambda \approx \pm \frac{HN}{(c - U_0)} \tag{10.12}$$

$$\lambda^{-1/2} \approx \left| \frac{(c - U_0)}{HN} \right|^{1/2}. \tag{10.13}$$

As we have already noted, the local VWL approaches zero as $(c - U_0) \to 0$. At the same time, w^* also $\to 0$ as $(c - U_0) \to 0$.

What about other fields? From (10.6) we have

$$u' = \frac{i}{k} \left(\frac{dw^*}{dz^*} - w^* \right). \tag{10.14}$$

---

[1] Our approach here follows Lindzen (1981).

From (10.7) we have

$$T' = -\frac{i}{k}\frac{H^2 N^2}{c - U_0} w^*.$$ (10.15)

From (10.5)

$$\Phi' = \frac{i}{k}(c - U_0)\left(\frac{dw^*}{dz^*} - w^*\right) + w^*\frac{dU_0}{dz^*}\frac{i}{k}.$$ (10.16)

We also want w in terms of w^*. Now

$$w^* = -\frac{1}{p_0}\left(\underbrace{\frac{\partial p'}{\partial t} + U_0\frac{\partial p'}{\partial x}}_{ik(U_0-c)p'=ik p_0(U_0-c)\Phi'} + w\underbrace{\frac{dp_0}{dz}}_{-\frac{p_0}{H}w}\right)$$

(*viz.* Chapters 4 and 9).

So, using (10.16),

$$-p_0 w^* = p_0(c - U_0)\left\{(c - U_0)\left(\frac{dw^*}{dz^*} - w^*\right) + w^*\frac{dU_0}{dz^*}\right\} - \frac{p_0}{H}w$$

or

$$w = Hw^* + \frac{1}{g}(c - U_0)\left\{(c - U_0)\left(\frac{dw^*}{dz^*} - w^*\right) + w^*\frac{dU_0}{dz^*}\right\}.$$ (10.17)

Assuming large λ^2 again

$$u' \approx \frac{i}{k}(-i\lambda)w^{*'} \approx \frac{A}{k}\left|\frac{HN}{(c - U_0)}\right|^{1/2} e^{z^*/2}e^{-i\int^{z^*}\lambda dz^*}$$ (10.18)

$$T' \approx -\frac{i}{k}\frac{H^2 N^2}{|c - U_0|^{1/2}}\frac{A}{|HN|^{1/2}}e^{z^*/2}e^{-i\int^{z^*}\lambda dz^*}$$ (10.19)

$$\Phi' \approx \frac{i}{k}(c - U_0)(-i\lambda)w^* \approx \frac{HN}{k}Ae^{z^*/2}\left|\frac{c - U_0}{HN}\right|^{1/2}e^{-i\int^{z^*}\lambda dz^*}$$ (10.20)

$$w' \approx Hw^* \approx AH\left|\frac{c - U_0}{HN}\right|^{1/2}e^{z^*/2}e^{-i\int^{z^*}\lambda dz^*}.$$ (10.21)

Thus, while w' and Φ' go to zero as $c - U_0 \to 0, u'$ and T' blow up. Is this consistent with the second Eliassen-Palm theorem?

$$\underbrace{\rho_0 \overline{u'w'}}_{p_0 = \rho_0 g H} = \frac{p_0}{gH} \frac{1}{2} |u'||w'|$$

since (10.18) and (10.21) imply u' and w' are in phase, and $\frac{1}{\pi} \int_0^\pi \sin^2 \phi \, d\phi = \frac{1}{2}$. From (10.18) and (10.19) we have

$$\rho_0 \overline{u'w'} = \frac{p_0(0)e^{-z^*}}{gH} \frac{1}{2} \frac{A}{k} e^{z^*/2} \cdot AH e^{z^*/2}$$

$$= \frac{1}{2} \frac{A^2}{gk} p_0(0) = \text{constant}.$$

The answer, therefore, is yes. It is interesting to note that the factor $\lambda^{-1/2}$ in Equation 10.10 is crucial in this regard. The momentum flux $\rho_0 \overline{u'w'}$ is actually the Wronskian of the wave equation, and the factor $\lambda^{-1/2}$ guarantees that the Wronskian of the WBKJ solution remains constant.

10.2   Critical level behaviour

We may next ask what happens when $c - U_0$ actually goes through zero. Clearly something must happen. Eliassen and Palm's first theorem tells us that if a wave travels through a critical layer, $\rho_0 \overline{u'w'}$ must change sign. This implies an exchange of momentum flux carried by the wave with the mean flow at the critical level. The only possible alternative is that the wave is totally reflected at the critical level, in which case $\overline{p'w'} = \rho_0 \overline{u'w'} = 0$.

An answer, for linear theory in the limit of vanishing damping, was obtained by Booker and Bretherton (1967). Let $z^* = 0$ be the critical level and let

$$U_0 = c + \frac{dU_0}{dz^*} z^*$$

in the neighbourhood of $z^* = 0$. Equation 10.8 becomes

$$\frac{d^2\tilde{w}}{dz^{*2}} + \left\{ \underbrace{\frac{H^2N^2}{(\frac{dU_0}{dz^*})^2 z^{*2}}}_{this\ term\ dominates} + \frac{\frac{dU_0}{dz^*}}{-(\frac{dU_0}{dz^*})z^*} - \frac{1}{4} \right\} \tilde{w} = 0,$$

which, in the neighbourhood of $z^* = 0$, is approximated by

$$\frac{d^2\tilde{w}}{dz^{*2}} + \frac{H^2N^2}{(\frac{dU_0}{dz^*})^2 z^{*2}}\tilde{w} = 0. \tag{10.22}$$

Equation 10.22 is simply a special case of Euler's equation, whose solution is

$$\tilde{w} = Az^{*\nu}, \tag{10.23}$$

where

$$\nu(\nu - 1) + \underbrace{\frac{H^2N^2}{(\frac{dU_0}{dz^*})^2}}_{=\frac{N^2}{(\frac{dU_0}{dz})^2} \equiv Ri} = 0 \tag{10.24}$$

or

$$\nu = \frac{1 \pm \sqrt{1 - 4Ri}}{2}. \tag{10.25}$$

10.2.1 Richardson number

The non-dimensional number, Ri, is known as the *Richardson number*. The nature of our solution will depend greatly on whether $Ri \lessgtr 1/4$. Now Ri is a measure of how rapidly $(c - U_0)$ is varying and for WKBJ theory to be appropriate we would want $Ri > 1/4$. In the atmosphere Ri is typically 1–10. In this case (10.23) becomes

$$\tilde{w} = |z^*|^{1/2} e^{\pm i\mu \ln z^*}, \tag{10.26}$$

where

$$\mu = \sqrt{Ri - 1/4},$$

and the + sign is appropriate to upward propagation (show this for yourself). Note that (10.26) is essentially our WKBJ solution in the neighbourhood of $z^* = 0$. Note also that this is very definitely not the case when $Ri < 1/4$! (It is not irrelevant to note that when $Ri < 1/4$ the fluid can be and often is unstable; i.e., waves draw energy from the basic state.)

10.2.2   Conditions for absorption

For $z^* > 0$, (10.26) becomes

$$\tilde{w} = z^{*1/2}e^{i\mu \ln z^*}. \tag{10.27}$$

Equation 10.27 has a branch point at $z^* = 0$. In continuing z^* to negative values, (10.27) does not lead to a unique answer. The answer depends on whether we pass above or below $z^* = 0$ in traversing the complex z^*-plane.[2]

---

[2]For those who don't remember what a branch point is, here are a few explanatory remarks.

Consider z^* in the complex plane

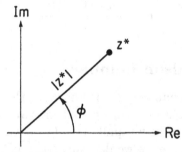

We may write $z^* =| z^* | e^{i\phi}$. Then

$$\begin{aligned}\ln z^* &= \ln(|z^*|e^{i\phi}) \\ &= \ln|z^*| + \ln e^{i\phi} \\ &= \ln|z^*| + i\phi.\end{aligned}$$

Now, ϕ in the above diagram is arbitrary to a positive or negative multiple of 2π. The choice of branch is tantamount to the choice of the appropriate 2π range for ϕ. If we are moving across the origin along the real axis, we must, in this connection,

Now, had we retained damping

$$-ik(c - U_0) \rightarrow \quad - \quad ik(c - U_0) + a$$

$$= -ik(\tfrac{ai}{k} - \tfrac{dU_0}{dz^*} z^*).$$

The branch point is now

$$z^* = \frac{\tfrac{ai}{k}}{\tfrac{dU_0}{dz^*}}, \ \text{Im} \ z^* > 0.$$

So with damping the branch point moves above $z^* = 0$, and hence in going from positive to negative z^* we pass below the branch point; that is, z^* goes to $|z^*|e^{-i\pi}$, and (10.27) goes to

$$\tilde{w} = -i|z^*|^{1/2}e^{i\mu(\ln|z^*| - i\pi)} = -i|z^*|^{1/2}e^{\mu\pi}e^{i\mu\ln|z^*|}. \quad (10.28)$$

It is easily shown from (10.27) and (10.28) that if $\rho_0\overline{u'w'} = A$ below $z^* = 0$, it will equal $-Ae^{-2\mu\pi}$ above. For $Ri > 1$, this implies almost complete *wave absorption* at the critical level. This is consistent with our earlier observation that group velocity $\rightarrow 0$ as $c - U_0 \rightarrow 0$, thus allowing any damping enough time to act.

10.2.3 Linear and nonlinear limits

By now you may be wondering whether we may not be pushing matters too far: solutions which are blowing up and finite momentum fluxes deposited in infinitesimal layers. At the very least, solutions blowing up might suggest that linearity is in question. For solutions in the neighbourhood of the critical level let us examine the relative magnitudes of linear and nonlinear terms. We have

$$w^* \approx A|z^*|^{1/2}e^{i\mu\ln|z^*|} \qquad \underbrace{e^{z^*/2}}_{\text{incorporate this into } A}$$

---

decide whether ϕ is going from 0 on the positive side to π on the negative side or from 0 on the positive side to $-\pi$ on the negative side. The former corresponds to passing above the origin; the latter to passing below the origin.

Let us, for simplicity, assume $\mu \gg 1$. Then

$$\frac{dw^*}{dz^*} \approx A\sqrt{-z^*}\left(\frac{-i\mu}{z^*}\right)e^{i\mu \ln(-z^*)}.$$

Also

$$iku' \approx -\frac{dw^*}{dz^*}$$

$$u' \approx \frac{i}{k}A\sqrt{-z^*}\left(\frac{-i\mu}{z^*}\right)e^{i\mu \ln(-z^*)}$$

$$\approx \frac{A\mu}{k}\frac{1}{\sqrt{-z^*}}e^{i\mu \ln(-z^*)}$$

and

$$\frac{du'}{dz^*} \approx \frac{A\mu}{k}\frac{1}{\sqrt{-z^*}}\left(\frac{-i\mu}{z^*}\right)e^{i\mu \ln(-z^*)}$$

$$\approx \frac{Ai\mu^2}{k}\frac{1}{(-z^*)^{3/2}}e^{i\mu \ln(-z^*)}.$$

Typical nonlinear terms in the horizontal momentum equation are $u'\frac{\partial u'}{\partial x}$ and $w'\frac{\partial u'}{\partial z}$.

$$\left|u'\frac{\partial u'}{\partial x}\right| \approx \frac{A^2\mu^2}{k}\frac{1}{|z^*|}$$

$$\left|w^*\frac{\partial u'}{\partial z^*}\right| \approx \frac{A^2\mu^2}{k}\frac{1}{|z^*|}.$$

Typical linear terms are $ik(U_0 - c)u'$ and $w^*\frac{dU_0}{dz^*}$.

$$\left|w^*\frac{dU_0}{dz^*}\right| \approx A|z^*|^{1/2}\frac{dU_0}{dz^*}$$

$$\left|ik(U_0 - c)u'\right| \approx A\mu\frac{dU_0}{dz^*}|z^*|^{1/2}.$$

(Note the latter is bigger.) The ratio of linear to nonlinear terms is

$$\left| \frac{\text{linear}}{\text{nonlinear}} \right| \approx \left| \frac{A\mu \frac{dU_0}{dz^*} |z^*|^{1/2}}{\frac{A^2\mu^2}{k}|z^*|^{-1}} \right| \approx \frac{k\frac{dU_0}{dz^*}}{\mu A}(-z^*)^{3/2}.$$

At that z^* for which this ratio ≈ 1 we expect nonlinear effects to be important. Now $\mu \sim Ri^{1/2}(Ri = RS/(\frac{dU_0}{dz^*})^2$, where $S = \frac{dT_0}{dz^*} + \kappa T_0)$. The ratio ≈ 1, when

$$(-z_{NL}^*) \sim \left(\frac{RiA}{k(RS)^{1/2}}\right)^{2/3}. \tag{10.29}$$

Not surprisingly z_{NL}^* depends on the wave amplitude. Now nonlinear problems are generally difficult. Can damping help us avoid these difficulties (and, at the same time, preserve our earlier linear results on what happens at a critical level)?

If we keep the linear damping shown in Equations 10.1 and 10.4, it is equivalent to replacing $ik(U_0 - c)$ with $ik(U_0 - c) + a$, or replacing $(c - U_0)$ with $(c - U_0) + \frac{ai}{k}$. Damping will begin to dominate when

$$\frac{a}{k} \sim (c - U_0)$$

or

$$\frac{a}{k} \sim \left| \frac{dU_0}{dz^*}z^* \right|$$

or when

$$z^* = z_d^* \approx -\frac{a}{k\frac{dU_0}{dz^*}} \approx -\frac{a}{k}\left(\frac{Ri}{RS}\right)^{1/2}. \tag{10.30}$$

When $|z_d^*| > |z_{nl}^*|$, nonlinearity will not have a chance to dominate and our earlier linear analysis will be appropriate – except that wave absorption will take place over a finite layer.

10.3   Damping and momentum deposition

The effect of damping is really more extensive than the above argument suggests. In the presence of damping, (10.12) becomes

$$\lambda \approx \frac{HN}{(c - U_0) + \frac{ai}{k}} \approx \frac{NH(c - U_0)}{(c - U_0)^2 + \frac{a^2}{k^2}} - \frac{NH(\frac{a}{k})i}{(c - U_0)^2 + \frac{a^2}{k^2}}$$

$$\approx \lambda_r + i\lambda_i. \tag{10.31}$$

λ now has an imaginary component which produces exponential decay – in addition to propagation. (N.B. we have decay because of our choice of a minus sign in (10.10).) This decay means that for a given wave forcing below the critical level, the effective values of A (*viz.*, (10.29)) near the critical level will be diminished – and the chances of damping dominating are increased.

How these matters actually work out is still a matter of some controversy. For a problem we will soon look at, the gravity wave periods will be rather long and radiative cooling will provide damping sufficient to obviate nonlinear effects. (In this case damping appears in (10.4), but not in (10.1). As an exercise you will show that this does not qualitatively alter the above results.) Other cases are not so clear. Benney and Bergeron (1969) have shown that the nonlinear limit leads to reflection rather than absorption at the critical level. However, the nonlinear solution is unstable and may lead to turbulence which in turn may lead to absorption again.

10.3.1   Violation of the second Eliassen-Palm theorem

We now come to an obvious – but major – point. When we have damping, the second Eliassen-Palm theorem is violated. Instead, we have

$$\rho_0 \overline{u'w'} = (\rho_0 \overline{u'w'})_0 e^{-2 \int_0^{z^*} \lambda_i dz^*}, \tag{10.32}$$

where $z^* = 0$ is assumed not to be a critical level, and

$$\frac{d}{dz}\rho_0\overline{u'w'} = \frac{1}{H}\frac{d}{dz^*}\rho_0\overline{u'w'} = \frac{-2\lambda_i}{H}\rho_0\overline{u'w'}. \tag{10.33}$$

The response of the mean flow (in the absence of rotation) will be given by

$$\frac{\partial\bar{u}}{\partial t} - \underbrace{\nu}_{\text{viscosity}}\frac{\partial^2\bar{u}}{\partial z^2} = -\frac{2\lambda_i}{H\rho_0}(\rho_0\overline{u'w'})_0 e^{-2\int_0^{z^*}\lambda_i dz^*}, \tag{10.34}$$

that is, wave absorption will lead to mean flow acceleration. It is also important to note that absorption increases markedly as $\bar{u}$ approaches c. (It is worth noting that in the presence of rotation, the response of $\bar{u}$ is no longer so clear. The Coriolis term can balance part of the Reynolds stress divergence, and the change in $\bar{u}$ might be negligible.) This mechanism is believed to play a major role in generating the quasi-biennial oscillation of the tropical stratosphere.

10.4 Quasi-biennial oscillation

Let us briefly recall what this phenomenon is. Its main features are shown in Figure 5.17. Easterly and westerly regimes in the equatorial stratosphere (of magnitude 20 m/s) alternate quasi-periodically with an average period of about twenty-six months. There is also an apparent downward phase propagation. Now, how do gravity waves enter this picture? Remember, for vertical propagation to be possible gravity waves must have periods short compared to the pendulum day. For most of the earth this means periods shorter than a day or so, and from what we saw in Chapter 5, most eddy energy is at periods longer than this. So, over most of the earth we do not anticipate that the bulk of the eddies will behave as gravity waves. However, as we approach the equator, the pendulum day goes to infinity, and the energy in these long periods can begin to propagate vertically as internal gravity waves. Also, the Coriolis term legitimately drops out of (10.34). It is, in fact, observed that such waves are produced in the neighbourhood of the equator – with both easterly and westerly phase speeds. The exact nature of their excitation is not yet fully understood, but it appears

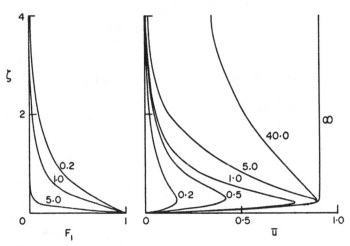

Figure 10.1: Profiles of mean velocity $\bar{u}$ and wave momentum flux F for single-wave forcing. Curves are labelled with time τ.

to originate near the tropical tropopause (*ca.* 16 km). Let us take the existence of such waves as given.

Figure 10.1 schematically illustrates how a fluid at rest (initially) might respond to a wave with phase speed $c = 1$. $\bar{u}$ is accelerated towards 1, but note that as $\bar{u}$ is brought towards 1 in the lower part of the domain, λ_i must increase and this prevents the wave from acting on the upper part of the domain. In the calculations upon which Figure 10.1 is based, diffusion eventually carries momentum upward. By the time a sharp shear zone develops near the bottom, the fluid above this shear zone is pretty much isolated from the westerly wave. If, however, there were also an easterly wave with $c = -1$, it would be free to propagate upward accelerating $\bar{u}$ towards -1. The downward descent of the easterly shear zone (on top of the previous westerly shear zone) would produce severe stresses which would wipe out the westerly zone and allow the westerly waves to propagate upward again starting the whole process anew. Of course, the actual calculations have some nuances which are here omitted. But, the essential mechanism is as described. In particular, the *amplitude* of the oscillation in $\bar{u}$ is determined by the *phase speeds* of the upward propagating waves, and the *period* of the oscillation in $\bar{u}$ is determined by the *average intensity* of the upward propagating waves.

This mechanism was first described by Lindzen and Holton (1968), and brought to its current form by Lindzen (1971) and Holton and Lindzen (1972). A highly simplified version of the theory appropriate to a laboratory configuration was developed by Plumb (1977) and the mechanism has, in fact, been simulated in the laboratory by Plumb and McEwan (1978).

Exercises

10.1 Rederive the vertical structure equation when there is linear damping in Equation 10.4 but not in Equation 10.1. Find the imaginary part of λ and compare with Equation 10.31. Note that thermal damping alone can lead to momentum flux deposition in the mean flow.

The following problems call for the use of the numerical gravity wave model described in Appendix A.

10.2 *WKBJ solution versus numerical solution*

Consider the following zonal wind profile

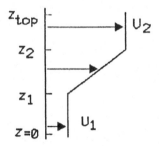

with the basic state temperature constant. Investigate the WKBJ versus numerical solution for various choices of the Richardson number (recall $Ri = N^2/U_{0z}^2$) and damping.

The following are some suggested parameter choices: number of levels = 500; $\tilde{w}_{bot} = 1$; $c_r = -20$ m/s; $c_i = 0$; $T_0 = 300°$K; $z_{top} = 30$ km; $z_1 = 10$ km; $z_2 = 20$ km; and $U_1 = -10$ m/s. Set interior forcing to zero, and use the radiation condition at the top. Choose values of U_2 that will correspond to Richardson numbers of 0.4, 1.0, and 5.0.

10.3 *Numerical resolution of critical lines*

Specify the phase velocity (c_r) such that there is a critical level somewhere in the domain. Investigate the numerical solution for various choices of the damping rate and number of levels. It is suggested that you try 1000 levels, with the profile and boundary conditions in Exercise 10.2 that corresponded to $Ri = 5$. Try choices of c_i that do and don't permit damping to dominate over the numerically resolved distance (*viz.* Equation 10.30).

Chapter 11

Rossby waves and the Gulf Stream

Supplemental reading:

Pedlosky[1] (1979), sections 3.1–3.3, 5.1–5.3, 5.5

11.1   Shallow water equations

In Chapter 9 we considered the general problem of linearized oscillations in a static, arbitrarily stratified atmosphere or ocean. Due to separability of latitude and altitude dependence, our equations naturally divide into two sets.

The first set consists in the following three equations:

$$i\sigma u'_n - f v'_n = -ik\Phi'_n \tag{11.1}$$

$$i\sigma v'_n + f u'_n = -\frac{\partial \Phi'_n}{\partial y} \tag{11.2}$$

and

$$iku'_n + \frac{\partial v'_n}{\partial y} = -\frac{i\sigma}{gh_n}\Phi'_n. \tag{11.3}$$

---

[1]Pedlosky offers very detailed explorations of the topics we are skimming. Don't get too discouraged in reading this book. It was the basis for a five trimester course at the University of Chicago

The second set consists in the following two equations:

$$-\frac{i\sigma}{gh_n}\Phi'_n + \frac{dw_n^*}{dz^*} - w_n^* = 0 \qquad (11.4)$$

and

$$i\sigma\frac{d\Phi'_n}{dz^*} + w_n^* R\left(\frac{dT_0}{dz^*} + \frac{RT_0}{c_p}\right) = \kappa J_n. \qquad (11.5)$$

The second set determines vertical structure – leading to a *vertical structure equation*

$$\frac{d^2\tilde{w}_n}{dz^{*2}} + \left\{\frac{1}{h_n}\left(\frac{dH}{dz^*} + \kappa H\right) - \frac{1}{4}\right\}\tilde{w}_n = \frac{\kappa J_n}{gh_n}e^{-z^*/2}, \qquad (11.6)$$

where

$$w_n^* = e^{z^*/2}\tilde{w}_n. \qquad (11.7)$$

At a horizontal rigid lower boundary, the boundary condition on $\tilde{w}_n$ is

$$\frac{d\tilde{w}_n}{dz^*} + \left(\frac{H}{h_n} - \frac{1}{2}\right)\tilde{w}_n = 0 \text{ at } z^* = 0. \qquad (11.8)$$

At a free upper surface we would require

$$\tilde{w}_n = 0 \text{ at } z^* = H^*. \qquad (11.9)$$

For an unbounded atmosphere we require boundedness and/or the radiation condition.

In Chapter 9 we outlined our procedure for forced oscillations. We now turn to the procedure for free oscillations. For free oscillations we solve (11.6) subject to the boundary conditions for $J = 0$. This yields the equivalent depths of the atmosphere (or ocean). For these depths we then solve (11.1)–(11.3) where frequency rather than h is the eigenvalue. The resulting modes are known as free oscillations.

As an exercise, you have shown that for a semi-infinite atmosphere with an isothermal basic state there exists a single equivalent depth for the atmosphere

$$h = \gamma H. \qquad (11.10)$$

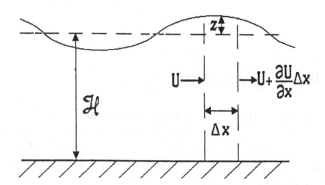

Figure 11.1: Shallow water geometry.

We have also shown that for a shallow fluid with a free surface and $N^2 = 0$ (i.e., $\frac{dH}{dz^*} + \kappa H \equiv 0$) there is also a single equivalent depth

$$h = \mathcal{H}^* H = \mathcal{H} = \text{depth of unperturbed fluid.} \qquad (11.11)$$

Under the latter circumstances, (11.1)–(11.3) are known as the *shallow water equations*. The point is that for a shallow fluid u' and v' are essentially independent of z, and Φ is just $g\times$ surface displacement. Equation 11.3 can then be directly interpreted as the continuity equation. This is readily seen if we ignore y-variation. Referring to Figure 11.1 we see that as $\Delta x \to 0$, $\mathcal{H}\frac{\partial u'}{\partial x} + \frac{\partial z'}{\partial t} = 0$. Under these circumstances there is no further need for the vertical structure equation. What our analysis shows is that the shallow water equations are always appropriate to hydrostatic oscillations on a static basic state – subject to the reinterpretation of h. Our analysis, therefore, serves to justify and generalize the use of the shallow water equations.

The shallow water equations are used as model equations in both meteorology and oceanography. In most applications they are not considered approximations. We shall make extensive use of these equations in the following two subsections where we investigate the effects of f, the Coriolis parameter, varying with y.

It should be remembered, however, that once we obtain a relation among frequency, horizontal wavenumbers, and the depth of the 'shallow water', we are always free to reinterpret the depth as an equivalent depth – and to solve for the depth in terms of the other parameters in order to apply the results to problems of forced rather than free oscillations.

11.2 Rossby waves

In Chapter 9 we saw that when f was constant and (11.1)–(11.3) were applied to a channel geometry, they described easterly and westerly gravity waves modified by rotation. We will now allow f to vary in the same way that density varies in a Boussinesq fluid; that is, we will take f to be constant unless it is differentiated – in which case we will take $\frac{df}{dy} = \beta = $ constant. This is known as the β-plane approximation and is, in fact, due to Rossby. Rossby intuitively realized that for large-scale meteorological systems the first modification on a sphere to the f-plane equations arises from the variation of f with latitude. Subsequently, scaling analysis has been used to more rigorously justify this approximation.

Our starting point will be (11.1)–(11.3). With variable f, it will prove more convenient to reduce (11.1)–(11.3) to a single equation in v' rather than Φ'. For free oscillations it doesn't matter which field we solve for. Eliminating u' between (11.1) and (11.2) we get

$$(f^2 - \sigma^2)v' = -i\sigma\frac{\partial \Phi'}{\partial y} + ikf\Phi'. \tag{11.12}$$

Similarly, eliminating u' between (11.1) and (11.3) yields:

$$i\sigma\frac{\partial v'}{\partial y} + ikfv' = \frac{\sigma^2}{gh}\Phi' - k^2\Phi'. \tag{11.13}$$

We next operate on (11.13) with $(-i\sigma\frac{\partial}{\partial y} + ikf)$ to get

$$\left(-i\sigma\frac{\partial}{\partial y} + ikf\right)\left(i\sigma\frac{\partial}{\partial y} + ikf\right)v' = \left(\frac{\sigma^2}{gh} - k^2\right)\left(-i\sigma\frac{\partial}{\partial y} + ikf\right)\Phi',$$

and using (11.12) we get

$$\left(-i\sigma\frac{\partial}{\partial y} + ikf\right)\left(i\sigma\frac{\partial}{\partial y} + ikf\right)v' = \left(\frac{\sigma^2}{gh} - k^2\right)\left(f^2 - \sigma^2\right)v',$$

or

$$\left\{\sigma^2\frac{\partial^2}{\partial y^2} \underbrace{+\sigma k f\frac{\partial}{\partial y}}_{1} +\sigma k\beta \underbrace{-\sigma k f\frac{\partial}{\partial y}}_{1} \underbrace{- k^2 f^2}_{2}\right\} v'$$

$$= \left\{\frac{\sigma^2 f^2}{gh} - \frac{\sigma^4}{gh} \underbrace{- k^2 f^2}_{2} + k^2\sigma^2\right\} v',$$

where the terms with numbered underbraces cancel. Finally, (dividing by σ^2) we get

$$\frac{\partial^2 v'}{\partial y^2} + \left(\frac{k\beta}{\sigma} + \frac{\sigma^2 - f^2}{gh} - k^2\right)v' = 0. \tag{11.14}$$

(It turns out that we also have Kelvin wave solutions for which $v' \equiv 0$. These are described at the end of this chapter.) If we again adopt a channel geometry where

$$v' = 0 \text{ at } y = 0, d, \tag{11.15}$$

then (11.14) has solutions

$$v' = \sin\frac{n\pi y}{d},$$

and (11.14) becomes

$$-\left(\frac{n\pi}{d}\right)^2 + \left(\frac{k\beta}{\sigma} + \frac{\sigma^2 - f^2}{gh} - k^2\right) = 0. \tag{11.16}$$

Recall that for free oscillations, h is given, and (11.16) is solved for σ, whereas in a forced problem, σ and k are given and (11.16) is solved for h. Although we shall return to free oscillations shortly, some useful insights can be gotten by looking at (11.16) as an equation for h:

$$gh = \frac{\sigma^2 - f^2}{(\ell^2 + k^2) - \frac{k\beta}{\sigma}} = \frac{\sigma^2 - f^2}{k^2\{\frac{\ell^2}{k^2} + 1 - \frac{\beta}{\sigma k}\}}, \qquad (11.17)$$

where $\ell = \frac{n\pi}{d}$.

In Chapter 9 we found (in the absence of β) that oscillations for which $\sigma^2 < f^2$ could not propagate vertically (propagation required $0 < h \lesssim H$; for $\sigma^2 < f^2$, h was negative). When β is included, (11.17) shows that this is no longer strictly true. For easterly waves (where σ and k have the same sign) and sufficiently *small* σ, vertical propagation is again possible. The resulting waves are called internal Rossby waves. Some quantitative estimates will help us determine the relevant scales for such waves. Let us take our β-plane to be centered at 45° latitude. Then

$$f = 2\Omega \sin\phi \approx 10^{-4}\,\text{sec}^{-1}$$

$$\beta = \frac{df}{dy} = \frac{2\Omega \cos\phi}{a} \approx \frac{10^{-4}\,\text{sec}^{-1}}{a},$$

where a = Earth's radius ≈ 6400 km.

For convenience let's take $n = 1$ and $d = a$. Then

$$\ell^2 \approx \left(\frac{\pi}{a}\right)^2.$$

Also

$$k = \frac{s}{a\cos\phi} = \frac{s\sqrt{2}}{a}$$

and

$$k^2 = \frac{2s^2}{a^2}.$$

From (11.17) we see that positive equivalent depths are possible for

$$0 < \sigma < \frac{\beta/k}{1 + \frac{\ell^2}{k^2}} = \frac{f/\sqrt{2}s}{1 + \frac{1}{2}(\frac{\pi}{s})^2} \le \frac{f}{2\pi}. \qquad (11.18)$$

(The reader should derive the last inequality.) In other words, Rossby waves are associated with periods for which $\sigma^2 \ll f^2$.

11.2.1 Planetary scale internal stationary waves

There is a quick application of the above. Just as flow over individual mountains is a major forcing of internal gravity waves, so too is flow over larger-scale surface features (the Tibetan Plateau, for example) a major forcing of internal Rossby waves. Now as we saw in Chapter 10 the inclusion of a basic flow can substantially complicate matters (indeed, if U_0 varies with z then, at least on a rotating sphere, we lose separability). However, in the trivial instance where $U_0 = $ constant, the primary effect is only to replace σ with $\sigma + kU_0$. (The effect on the lower boundary condition can be more complicated – but we will ignore this.) Now for stationary waves $\sigma = 0$ and (11.18) becomes

$$0 < kU_0 < \frac{f/\sqrt{2}s}{1 + \frac{1}{2}(\frac{\pi}{s})^2}$$

or

$$0 < U_0 < \frac{fa}{2s^2(1 + \frac{1}{2}(\frac{\pi}{s})^2)} = \frac{fa/2}{s^2 + \frac{\pi^2}{2}} \equiv U_{trap}, \qquad (11.19)$$

where

$$U_{trap} = \begin{cases} 54\text{m/s for } s = 1 \\ 36\text{m/s for } s = 2 \\ 23\text{m/s for } s = 3 \end{cases}$$

Only sufficiently weak westerlies permit stationary wave propagation. We see from (11.19) that for tropospheric winds stationary wavenumbers three and greater will not readily propagate into the stratosphere thus accounting for the predominance of wavenumbers one and two in the winter stratosphere. We also see that the summer stratospheric easterlies will effectively block the propagation of all stationary waves accounting for the observed zonal character of the summer stratospheric circulation. These results were first noted by Charney and Drazin (1961).

11.2.2   Free oscillations

Let us now return to free oscillations. Rewriting (11.16) we get

$$\frac{k\beta}{\sigma} + \frac{\sigma^2 - f^2}{gh} - \left(k^2 + \left(\frac{n\pi}{d}\right)^2 \right) = 0. \qquad (11.20)$$

Equation 11.20 is cubic in σ. We may anticipate that the three roots will correspond to two gravity waves and a Rossby wave. In general, gravity wave frequencies (for positive h) will exceed f so the term $\frac{k\beta}{\sigma}$ will be negligible. Similarly, for Rossby waves $\sigma^2 \ll f^2$. Thus (11.20) has the following approximate solutions:

$$\sigma^2 \approx gh \left(k^2 + \left(\frac{n\pi}{d}\right)^2 \right) + f^2 \qquad (11.21)$$

and

$$\frac{k\beta}{\sigma} = \frac{f^2}{gh} + k^2 + \left(\frac{n\pi}{d}\right)^2$$

or

$$\sigma = \frac{k\beta}{k^2 + (\frac{n\pi}{d})^2 + \frac{f^2}{gh}}. \qquad (11.22)$$

In contrast to gravity waves, which can have easterly and westerly phase speeds, Rossby waves always have easterly phase speed (relative to U_0). East–west asymmetry is always a characteristic of motions for which β is important.

Several properties of (11.22) are worth noting:

(i) σ has a maximum value for a particular value of k (which you will work out as an exercise). Pedlosky (1979) has an elegant explanation of this which we shall go over as soon as we develop some theorems on vorticity conservation.

(ii) With a constant zonal flow we can replace σ with $\sigma + kU_0$. Also $c = -\frac{\sigma}{k}$. (11.22) becomes

$$c = U_0 - \frac{\beta}{k^2 + (\frac{n\pi}{d})^2 + \frac{f^2}{gh}}. \qquad (11.23)$$

Note that as $(k^2 + (\frac{n\pi}{d})^2)$ gets larger the Rossby wave moves more and more nearly with the basic flow U_0. This is completely consistent with observations.

(iii) The velocity fields associated with Rossby waves are almost in geostrophic balance. Nonetheless, the outright assumption of geostrophy would not permit us to evaluate the time evolution of Rossby waves. The development of an appropriate approximate set of equations which exploit geostrophy but still describe Rossby waves will be one of our tasks.

(iv) The dispersive properties of Rossby waves clearly suggests that it is Rossby waves and not gravity waves which describe large-scale meteorological systems.

We next turn to a rather different application of the shallow water equations wherein we again show how β introduces a profound east–west asymmetry.

11.3 Intensification of ocean currents

Figure 11.2 shows a schematic description of the world's ocean currents. A characteristic of these currents is intensification at western boundaries. It is generally accepted that the main current systems in the ocean are forced by wind stresses. However, this does not, *per se*, account for westward intensification. Stommel (1948) first showed that the presence of β (in conjunction with friction in his model) leads to westward intensification. Although Stommel's model is not considered to be an accurate representation of the detailed physics, it is still the simplest model which displays the westward intensification, and we will therefore, concentrate on it. A very general discussion of the wind driven ocean circulations is given in chapter 5 of Pedlosky (1979).

Stommel's equations consist simply in the steady, linearized shallow water equations with vertical turbulent exchange on a β-plane:

$$- fv = \frac{\partial \Phi}{\partial x} + \frac{1}{\rho}\frac{\partial}{\partial z}\left(k_v \frac{\partial u}{\partial z}\right) \tag{11.24}$$

$$fu = -\frac{\partial \Phi}{\partial y} + \frac{1}{\rho}\frac{\partial}{\partial z}\left(k_v \frac{\partial v}{\partial z}\right) \tag{11.25}$$

and

$$\frac{\partial u}{\partial x} + \frac{\partial v}{\partial y} = 0, \tag{11.26}$$

where ρ is taken as constant. We next multiply (11.24)–(11.26) by ρ and integrate over the depth of the flat-bottomed ocean, H.

$$- fM_y = -H\frac{\partial \Phi}{\partial x} + \int_0^H \frac{\partial}{\partial z}\left(k_v \frac{\partial u}{\partial z}\right) dz \tag{11.27}$$

$$fM_x = -H\frac{\partial \Phi}{\partial y} + \int_0^H \frac{\partial}{\partial z}\left(k_v \frac{\partial v}{\partial z}\right) dz \tag{11.28}$$

and

$$\frac{\partial M_x}{\partial x} + \frac{\partial M_y}{\partial y} = 0, \tag{11.29}$$

where

$$M_x \equiv \int_0^H \rho u\, dz$$

$$M_y \equiv \int_0^H \rho v\, dz.$$

There are contributions to the integrals in (11.27) and (11.28) from both wind stress at the surface and from bottom drag.

$$\int_0^H \frac{\partial}{\partial z}\left(k_v \frac{\partial u}{\partial z}\right) dz = \tau_x - cM_x \tag{11.30}$$

$$\int_0^H \frac{\partial}{\partial z}\left(k_v \frac{\partial v}{\partial z}\right) dz = \tau_y - cM_y, \tag{11.31}$$

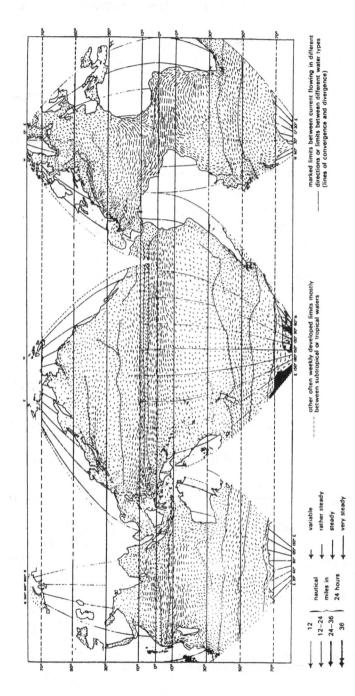

Figure 11.2: The circulation pattern of the world's oceans. This map (after Defant, 1961) represents a long-term compilation of measurements and ship reports. Although it is schematic rather than precise, it reveals the westward intensification of the circulation in the major ocean gyres.

where τ_x and τ_y are wind stress in the x and y directions and c is a bottom drag coefficient. (11.27) and (11.28) become

$$- fM_y = -H\frac{\partial \Phi}{\partial x} + \tau_x - cM_x \tag{11.32}$$

$$fM_x = -H\frac{\partial \Phi}{\partial y} + \tau_y - cM_Y. \tag{11.33}$$

To eliminate Φ we differentiate (11.32) with respect to y, differentiate (11.33) with respect to x, and subtract:

$$-fM_{y,y} - \beta M_y = -H\frac{\partial^2 \Phi}{\partial x \partial y} + \tau_{x,y} - cM_{x,y}$$

$$fM_{x,x} = -H\frac{\partial^2 \Phi}{\partial x \partial y} + \tau_{y,x} - cM_{y,x}$$

$$-\beta M_y = \tau_{x,y} - \tau_{y,x} - cM_{x,y} + cM_{y,x}. \tag{11.34}$$

From (11.29) we can write

$$M_x = -\frac{\partial \psi}{\partial y} \tag{11.35}$$

$$M_y = \frac{\partial \psi}{\partial x}, \tag{11.36}$$

where ψ is a stream function. With (11.35) and (11.36), (11.34) becomes

$$-\beta\frac{\partial \psi}{\partial x} = -\nabla \times \vec{\tau} + c\nabla^2 \psi. \tag{11.37}$$

The nature and magnitude of bottom friction are totally uncertain. However, it may be presumed to be very small since the curl of the wind stress is observed to be approximately balanced over the whole interior of the ocean (away from coasts) by the advection of the Earth's vorticity (Welander, 1959); that is, in the interior

$$\beta\frac{\partial \psi}{\partial x} \approx \nabla \times \vec{\tau}. \tag{11.38}$$

Equation 11.38 is known as the *Sverdrup relation*.

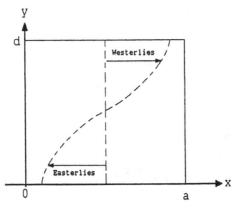

Figure 11.3: Geometry of ocean and wind stress in Stommel model.

For simplicity we will consider our ocean to be rectangular (*viz.* Figure 11.3). Our boundary conditions will simply be

$$\psi = 0 \text{ at } y = 0, d; x = 0, a. \tag{11.39}$$

(In the absence of lateral friction we can't use no-slip conditions.) We will assume the wind to be purely zonal:

$$\tau_y = 0$$

$$\tau_x = -T \cos \frac{\pi y}{d}. \tag{11.40}$$

Equation 11.37 becomes

$$-\beta \frac{\partial \psi}{\partial x} = \frac{\pi}{d} T \sin \frac{\pi y}{d} + c \nabla^2 \psi. \tag{11.41}$$

Let

$$\psi = f(x) \sin \frac{\pi y}{d}. \tag{11.42}$$

Equation 11.41 becomes

$$-\beta f_x = \frac{\pi}{d} T + c \left(f_{xx} - \left(\frac{\pi}{d} \right)^2 f \right). \tag{11.43}$$

Let us briefly examine the Sverdrup balance

$$-\beta f_x = \frac{\pi}{d}T$$

or

$$f_x = -\frac{\pi}{d}\frac{T}{\beta} \tag{11.44}$$

(i.e., we expect a southward drift everywhere in the interior). Integrating (11.44) we get

$$f = \frac{\pi}{d}\frac{T}{\beta}(-x + k),$$

where k is an integration constant, and

$$\psi_{\text{Sverdrup}} = \frac{\pi}{d}\frac{T}{\beta}(-x + k)\sin\frac{\pi y}{d}. \tag{11.45}$$

Equation 11.45 satisfies boundary conditions at $y = 0, d$, but it can at best satisfy either

$$\psi = 0 \text{ at } x = 0 \ (k = 0)$$

or

$$\psi = 0 \text{ at } x = a \ (k = a).$$

The consequences of either choice are illustrated in Figure 11.4. A situation where an interior solution cannot satisfy a boundary condition generally calls for a thin boundary layer where, although c is very small, $c f_{xx}$ will be able to balance βf_x. Because of β, virtually any added mechanism will cause the return flow to occur on the western boundary. The use of 'bottom friction' illustrates this process. First, let's scale x by $\frac{d}{\pi}$, and f by $\frac{T}{\beta}$. Equation 11.43 becomes

$$-\tilde{f}_{\tilde{x}} = 1 + \epsilon(\tilde{f}_{\tilde{x}\tilde{x}} - \tilde{f}), \tag{11.46}$$

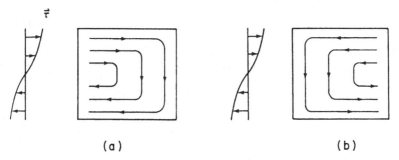

(a)   (b)

Figure 11.4: (a) The streamlines of the Sverdrup transport for the stress distribution shown at left of the interior geostrophic transport is made to satisfy the zero-normal-flow condition at the eastern boundary. Note the implied requirement of a *western* boundary current to complete the flow. (b) The Sverdrup streamlines for the *same* stress distribution as (a), where now the Sverdrup transport has been made to satisfy the no-normal-flow condition at the western boundary. The validity of this choice implicitly requires the existence of an eastern boundary current. After Pedlosky (1979).

where

$$f = \frac{T}{\beta}\tilde{f}$$

$$x = \frac{d}{\pi}\tilde{x}$$

$$\epsilon = \frac{c}{\beta}\left(\frac{\pi}{d}\right)$$

and

$$\tilde{f}(0) = \tilde{f}\left(\frac{\pi a}{d}\right) = 0.$$

Rewrite (11.46)

$$\tilde{f}_{\tilde{x}\tilde{x}} + \frac{1}{\epsilon}\tilde{f}_{\tilde{x}} - \tilde{f} = -\frac{1}{\epsilon}. \qquad (11.47)$$

Next let $\tilde{f} = \frac{1}{\epsilon} + g$:

$$g_{\tilde{x}\tilde{x}} + \frac{1}{\epsilon}g_{\tilde{x}} - g = 0. \tag{11.48}$$

Now,

$$g = ae^{\lambda_1\tilde{x}} + be^{\lambda_2\tilde{x}},$$

where

$$\lambda_{1,2} = \frac{-\frac{1}{\epsilon} \pm \sqrt{\frac{1}{\epsilon^2} + 4}}{2}$$

$$\lambda_1 \approx \epsilon$$

$$\lambda_2 \approx \frac{-1}{\epsilon}.$$

So,

$$\tilde{f} = \frac{1}{\epsilon} + ae^{\epsilon\tilde{x}} + be^{-\tilde{x}/\epsilon} \tag{11.49}$$

$$f(0) = \frac{1}{\epsilon} + a + b = 0 \tag{11.50}$$

and

$$f\left(\frac{\pi a}{d}\right) = \frac{1}{\epsilon} + ae^{\epsilon\pi a/d} + \underbrace{be^{-\pi a/\epsilon d}}_{negligible} = 0. \tag{11.51}$$

Equation 11.51 implies

$$a \approx -\frac{1}{\epsilon}e^{-\epsilon\frac{\pi a}{d}}.$$

Equation 11.50 implies

$$b \approx \frac{1}{\epsilon}(e^{-\epsilon\frac{\pi a}{d}} - 1);$$

and

$$\tilde{f} \approx -\frac{1}{\epsilon}(e^{\epsilon(\tilde{x}-\frac{\pi a}{d})} - 1) + \frac{1}{\epsilon}(e^{-\epsilon\frac{\pi a}{d}} - 1)e^{-\tilde{x}/\epsilon}$$

$$\approx -\left(\tilde{x} - \frac{\pi a}{d}\right) - \frac{\pi a}{d}e^{-\tilde{x}/\epsilon}. \tag{11.52}$$

Thus the interior solution satisfies the eastern boundary condition and the intense return flow occurs at the western boundary. The return flow is confined to a zone of width $\Delta\tilde{x} \sim \epsilon$.

11.4 Remark on Kelvin waves
The case of $v' \equiv 0$

The solutions in Section 2 assume $v' \neq 0$. We here consider the situation where $v' \equiv 0$. There exists a solution in this case known as a Kelvin wave. As an exercise you will derive the properties of Kelvin waves on what is known as an equatorial β -plane (where $f = \beta y$). Here we will take $f = f_0$. Equations 11.1–11.3 become

$$i\sigma u' = -ik\Phi' \tag{11.53}$$

$$fu' = -\frac{\partial \Phi'}{\partial y} \tag{11.54}$$

and

$$iku' = -\frac{i\sigma}{gH}\Phi'. \tag{11.55}$$

From (11.53) and (11.55) we get

$$\frac{\sigma^2}{k^2} = gH, \tag{11.56}$$

while from (11.53) and (11.54) we get

$$\frac{\partial \Phi'}{\partial y} - f_0 \frac{k}{\sigma}\Phi' = 0. \tag{11.57}$$

The solution to (11.57) is

$$\Phi' = Ce^{f_0(\frac{k}{\sigma})y}.$$

Note that for $k\sigma < 0$ (i.e., westerly or eastward propagating waves) amplitudes decay away from the boundary at $y = 0$, while for $k/\sigma > 0$ (i.e., easterly or westward propagating waves) amplitudes decay away from the boundary at $y = d$. In a closed basin, Kelvin waves travel around the boundary of the basin in a counterclockwise direction (when $f_0 > 0$). Kelvin waves satisfy the dispersion relation for a gravity wave in the absence of rotation; at the same time, the Kelvin wave's horizontal velocity field is in geostophic balance.

Exercises

11.1 Using a scale analysis, determine under what conditions the β-plane is an acceptable approximation to the rotating sphere.

11.2 Assess the accuracy of geostrophic balance for Rossby waves.

11.3 For the parameters used in connection with stationary waves (i.e., the values of $d, n = 0, k, f$, and β – and taking $h = 10.4$ km) find that value of k for which σ (for Rossby waves) is a maximum.

11.4 Write down the horizontal momentum equations for perturbations u', v', p' in a form applicable to an equatorial β-plane, that is, with $f = \beta y$. Assume a solution with $v' = 0$, and

$$u' = \text{Re}\{u_0' \exp i(wt - kx)\}.$$

Show that in this case the amplitude u_0' varies with the latitude as

$$\exp\left(\frac{-\beta y^2 k}{2\omega}\right)$$

Note that for a satisfactory solution k must be positive, that is, the wave must be eastward moving. Plot the pressure field associated with the variation in zonal velocity u'. These are equatorial Kelvin waves; they have been observed in the stratosphere.

Chapter 12

Vorticity and quasi-geostrophy

Supplemental reading:

Holton (1979), chapters 4, 6

Houghton (1977), sections 8.4–8.6

Pedlosky (1979), chapter 2, sections 3.10, 3.12, 3.13

12.1 Preliminary remarks

In the preceding chapter we saw that β plays a major role in large-scale motions of the atmosphere and ocean. We also referred to β as the gradient of that contribution to a fluid's *vorticity* due to the Earth's rotation. We will now briefly consider what exactly is vorticity.

Recall that in particle mechanics, conservation of energy and momentum both play important roles. In a fluid, however, momentum is not in general conserved because of the presence of pressure forces. To be sure, in symmetric circulations, zonal angular momentum is conserved (in the absence of friction), but then $\frac{\partial p'}{\partial x} = 0$ by definition. The question we will consider is whether there is anything a fluid conserves instead of momentum.

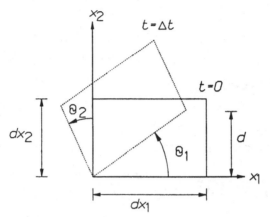

Figure 12.1: A rectangular element of fluid at $t = 0$ is deformed and rotated by the fluid flow into the rhomboidal element at $t = dt$.

12.1.1   Interpretation of vorticity

A clue is obtained from the following 'quasi-fluid' equation:

$$\frac{\partial \vec{u}}{\partial t} = -\frac{1}{\rho}\nabla p, \tag{12.1}$$

where $\rho = $ const. Taking the curl of (12.1) eliminates ∇p and leaves us with

$$\frac{\partial}{\partial t}(\nabla \times \vec{u}) = 0. \tag{12.2}$$

In this 'quasi-fluid' $\nabla \times \vec{u}$ could be considered as conserved. $\nabla \times \vec{u}$ is called vorticity. Vorticity can be interpreted as twice the instantaneous local rotation rate of an element of fluid. This is easily seen in two-dimensional flow. With reference to Figure 12.1,

$$d = \frac{\partial u_2}{\partial x_1}dx_1 dt$$

$$\Theta_1 = \frac{d}{dx_1} = \frac{\partial u_2}{\partial x_1}dt.$$

Similarly,

$$\Theta_2 = -\frac{\partial u_1}{\partial x_2}dt.$$

Now $\frac{d}{dt}(\Theta_1 + \Theta_2) \equiv 2\frac{d\alpha}{dt} = \frac{\partial u_2}{\partial x_1} - \frac{\partial u_1}{\partial x_2}$, which is what we set out to show.

The real equations of motion are more complicated than (12.1), but as we shall see a quantity closely related to vorticity is, in fact, conserved in inviscid, adiabatic fluids.

12.2 Vorticity in the shallow water equations

Let us first consider the shallow water equations introduced in Chapter 11. This time, however, we will consider the nonlinear shallow water equations. We will retain the β-plane geometry.

$$\frac{\partial u}{\partial t} + u\frac{\partial u}{\partial x} + v\frac{\partial u}{\partial y} - fv = -gZ_x \tag{12.3}$$

$$\frac{\partial v}{\partial t} + u\frac{\partial v}{\partial x} + v\frac{\partial v}{\partial y} + fu = -gZ_y \tag{12.4}$$

$$Z\nabla \cdot \vec{u} + \frac{DZ}{Dt} = 0. \tag{12.5}$$

To eliminate Z_x and Z_y in (12.3) and (12.4) (Z is the counterpart of pressure), we differentiate (12.3) with respect to y, and (12.4) with respect to x, and subtract the resulting equations:

$$\frac{\partial^2 u}{\partial t\partial y} + \frac{\partial u}{\partial y}\frac{\partial u}{\partial x} + u\frac{\partial^2 u}{\partial x\partial y} + \frac{\partial v}{\partial y}\frac{\partial u}{\partial y}$$

$$+ v\frac{\partial^2 u}{\partial y^2} - f\frac{\partial v}{\partial y} - \beta v = -g\frac{\partial^2 Z}{\partial x\partial y} \tag{12.6}$$

$$\frac{\partial^2 v}{\partial t \partial x} \quad + \quad \frac{\partial u}{\partial x}\frac{\partial v}{\partial x} + u\frac{\partial^2 v}{\partial x^2} + \frac{\partial v}{\partial x}\frac{\partial v}{\partial y}$$

$$+ \quad v\frac{\partial^2 v}{\partial x \partial y} + f\frac{\partial u}{\partial x} = -g\frac{\partial^2 Z}{\partial x \partial y} \tag{12.7}$$

$$\frac{\partial}{\partial t}\left(\frac{\partial v}{\partial x} - \frac{\partial u}{\partial y}\right) + u\frac{\partial}{\partial x}\left(\frac{\partial v}{\partial x} - \frac{\partial u}{\partial y}\right) + v\frac{\partial}{\partial y}\left(\frac{\partial v}{\partial x} - \frac{\partial u}{\partial y}\right)$$

$$+\beta v + f\left(\frac{\partial u}{\partial x} + \frac{\partial v}{\partial y}\right) + \frac{\partial u}{\partial x}\left(\frac{\partial v}{\partial x} - \frac{\partial u}{\partial y}\right)$$

$$+\frac{\partial v}{\partial y}\left(\frac{\partial v}{\partial x} - \frac{\partial u}{\partial y}\right) = 0 \tag{12.8}$$

or

$$\frac{D\zeta}{Dt} + \beta v + f\nabla \cdot \vec{u} + \zeta\nabla \cdot \vec{u} = 0, \tag{12.9}$$

where ζ = vertical component of vorticity = $\frac{\partial v}{\partial x} - \frac{\partial u}{\partial y}$.
Equation 12.9 may be rewritten

$$\frac{D}{Dt}(\zeta + f) + (\zeta + f)\nabla \cdot \vec{u} = 0. \tag{12.10}$$

The quantity $\zeta + f$ is called *absolute vorticity* while ζ is called *relative vorticity*. (Why ?) Using (12.5), (12.10) becomes

$$\frac{D}{Dt}(\zeta_a) - \frac{\zeta_a}{Z}\frac{DZ}{Dt} = 0$$

or

$$\frac{D}{Dt}\left(\frac{\zeta_a}{Z}\right) = 0, \tag{12.11}$$

where $\zeta_a = \zeta + f$. The quantity ζ_a/Z, known as *absolute potential vorticity*, is conserved by the shallow water equations. The existence of Rossby waves is closely related to the conservation of vorticity or potential vorticity. Recall that the shallow water equations allow both gravity waves and Rossby waves. If, however, we put a rigid lid on the fluid we will eliminate gravity waves. In such a situation,

$$\frac{DZ}{Dt} = 0$$

and (12.11) becomes

$$\frac{D\zeta_a}{Dt} = \frac{D}{Dt}\left(\frac{\partial v}{\partial x} - \frac{\partial u}{\partial y} + f\right) = 0. \tag{12.12}$$

Equation 12.5 becomes

$$\nabla \cdot \vec{u} = 0 \tag{12.13}$$

which implies the existence of a stream function such that

$$u = -\psi_y \tag{12.14}$$

$$v = \psi_x. \tag{12.15}$$

12.2.1 Filtered Rossby waves

If we assume a constant basic flow, u_o, and linearizable perturbations on this flow, Equation 12.12 becomes

$$\left(\frac{\partial}{\partial t} + u_0\frac{\partial}{\partial x}\right)(\nabla^2\psi + f) + \psi_x\beta = 0. \tag{12.16}$$

If we assume further that the perturbations are of the form

$$\sin \ell y \, e^{ik(x-ct)}$$

then (12.16) becomes

$$ik(u_0 - c)(-k^2 - \ell^2)\psi + ik\psi\beta = 0$$

or

$$c = u_0 - \frac{\beta}{k^2 + \ell^2}, \tag{12.17}$$

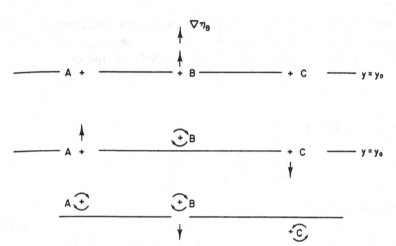

Figure 12.2: The position of the three-point vortices A, B and C at three successive times. Initially colinear and positioned along an isobar, B is displaced upwards, producing velocities at A and C which move them as shown. The vorticity induced on A and C produces a velocity at B tending to restore it to its original position. After Pedlosky (1979).

which is simply the equation for non-divergent Rossby waves. The mechanism of such waves is shown in Figure 12.2. Recall that ζ_a consists in both relative vorticity and f; f increases with y. Now if $\zeta = 0$ at $y = y_0$ and an element is displaced to a positive y, a negative ζ (clockwise rotation) will be induced to counteract the increasing f. The result will be a disturbance whose phase propagates westward relative to u_0.

We next consider what happens if we restore a free surface. If we linearize (12.11) about a constant u_0 basic state we get

$$\left(\frac{\partial}{\partial t} + u_0 \frac{\partial}{\partial x}\right)\left(\frac{\partial v'}{\partial x} - \frac{\partial u'}{\partial y}\right) + v'\beta - \frac{f}{Z_0}\left(\frac{\partial}{\partial t} + u_0 \frac{\partial}{\partial x}\right)Z' = 0. \quad (12.18)$$

Now the linearization of Equations 12.3 and 12.4 could be used to relate u' and v' to Z', but the resulting dispersion relation would be cubic in c. However, from the exercises we know that u' and v' in a Rossby wave are approximately geostrophic; that is,

$$v' \cong \frac{g}{f}Z'_x \quad (12.19)$$

and

$$u' \cong -\frac{g}{f} Z'_y. \tag{12.20}$$

Let's see what happens if we substitute (12.19) and (12.20) into (12.18):

$$\left(\frac{\partial}{\partial t} + u_0 \frac{\partial}{\partial x}\right)\left(\frac{g}{f}(Z'_{xx} + Z'_{yy})\right) + \frac{g}{f}\beta Z'_x - \frac{f}{Z_0}\left(\frac{\partial}{\partial t} + u_0 \frac{\partial}{\partial x}\right) Z' = 0. \tag{12.21}$$

Again, assuming solutions of the form

$$\sin \ell y \; e^{ik(x-ct)},$$

(12.21) becomes

$$-(u_0 - c)\frac{g}{f}(k^2 + \ell^2) + \frac{g\beta}{f} - \frac{f}{Z_0}(u_0 - c) = 0$$

or

$$-(u_0 - c)(k^2 + \ell^2) + \beta - \frac{f^2}{gZ_0}(u_0 - c) = 0$$

or

$$c = u_0 - \frac{\beta}{k^2 + \ell^2 + \frac{f^2}{gZ_0}}, \tag{12.22}$$

which is precisely the dispersion relation for divergent Rossby waves[1]. We seem to have found a way of exploiting geostrophy to suppress gravity waves while retaining the time evolution associated with Rossby waves. Our next step is to make this procedure systematic.

12.3   Quasi-geostrophic shallow water theory

Once one knows what one is after, scaling affords a convenient way to make things systematic. It will become transparently clear that if

---

[1]Note that in the non-divergent case, we automatically have a streamfunction, so that there is no need for a quasi-geostrophic approximation in order to obtain (12.17). Equivalently, the non-divergent case does not have surface gravity waves to filter out.

one does not know what one wants *a priori*, scaling is not nearly so effective!

Let us go through the ritual of scaling the dependent and independent variables in Equations 12.3–12.5 as follows:

$$
\begin{aligned}
u &= Uu' \\
v &= Uv' \\
x &= Lx' \\
y &= Ly' \\
Z &\equiv \underbrace{\bar{Z}}_{\text{mean depth}} + \tilde{Z}, \text{ and } \tilde{Z} = HZ' \\
f &= f + \beta y = f + \beta Ly' \\
t &= Tt'.
\end{aligned}
$$

In terms of dimensionless variables (12.3)–(12.5) become

$$
\frac{U}{T}\frac{\partial u'}{\partial t'} + \frac{U^2}{L}\left(u'\frac{\partial u'}{\partial x'} + v'\frac{\partial u'}{\partial y'}\right)
$$

$$
- fU\left(1 + \frac{\beta L}{f}y'\right)v' = -g\frac{H}{L}Z'_{x'}
$$

or

$$
\frac{1}{fT}\frac{\partial u'}{\partial t'} + \frac{U}{fL}\left(u'\frac{\partial u'}{\partial x'} + v'\frac{\partial u'}{\partial y'}\right) - \left(1 + \frac{\beta L}{f}y'\right)v' = -\frac{gH}{fUL}Z'. \quad (12.23)
$$

Similarly,

$$
\frac{1}{fT}\frac{\partial v'}{\partial t} + \frac{U}{fL}\left(u'\frac{\partial v'}{\partial x'} + v'\frac{\partial v'}{\partial y'}\right)
$$

$$
+ \left(1 + \frac{\beta L}{f}y'\right)u' = -\frac{gH}{fUL}Z'_{y'} \quad (12.24)
$$

and

$$
\left(1 + \frac{HZ'}{\bar{Z}}\right)\left(\frac{\partial u'}{\partial x'} + \frac{\partial v'}{\partial y'}\right) + \frac{HL}{\bar{Z}TU}\frac{\partial Z'}{\partial t'} + \frac{H}{\bar{Z}}\left(u'\frac{\partial Z'}{\partial x'} + v'\frac{\partial Z'}{\partial y'}\right) = 0. \quad (12.25)
$$

We wish to capitalize on the following to simplify our equations:

(i) The dominance of the Coriolis force;

(ii) The approximate validity of geostrophy;

(iii) The small excursions of f from its mean value.

Item (ii) leads to taking

$$fU = g\frac{H}{L}$$

or

$$H = \frac{fUL}{g}. \tag{12.26}$$

Item (i) leads to our taking

$$R_0 = \frac{U}{fL} \ll 1 \tag{12.27}$$

and

$$R_{0T} = \frac{1}{fT} \ll 1. \tag{12.28}$$

For *simplicity* we will take

$$R_0 = R_{0T} = \epsilon. \tag{12.29}$$

Item (iii) leads us to write

$$\frac{\beta L}{f} = \epsilon\beta'. \tag{12.30}$$

We may now rewrite (12.26) as

$$H = \epsilon\frac{f^2L^2}{g},$$

in which case the non-dimensional parameters in (12.25) become

$$\frac{H}{\tilde{Z}} = \epsilon\frac{f^2L^2}{g\tilde{Z}}$$

and

$$\frac{H}{\tilde{Z}}\frac{L}{TU} = \epsilon\frac{f^2L^2}{g\tilde{Z}}\frac{R_{0T}}{R_0} = \epsilon\frac{f^2L^2}{g\tilde{Z}}.$$

12.3.1 Rossby radius

Now let us define a distance R by the following relation

$$\frac{f^2 L^2}{g\bar{Z}} = \frac{L^2}{R^2}$$

or

$$R^2 = \frac{g\bar{Z}}{f^2};$$

R is known as the Rossby radius.

12.3.2 Rossby number expansion

We will take

$$L^2 = R^2. \tag{12.31}$$

As a result of the above, we may rewrite (12.23)–(12.25) as follows:

$$\epsilon \left(\frac{\partial u'}{\partial t'} + u' \frac{\partial u'}{\partial x'} + v' \frac{\partial u'}{\partial y'} \right) - (1 + \epsilon \beta' y') v' = -Z'_{x'} \tag{12.32}$$

$$\epsilon \left(\frac{\partial v'}{\partial t'} + u' \frac{\partial v'}{\partial x'} + v' \frac{\partial v'}{\partial y'} \right) - (1 + \epsilon \beta' y') u' = -Z'_{y'} \tag{12.33}$$

and

$$(1 + \epsilon Z') \left(\frac{\partial u'}{\partial x'} + \frac{\partial v'}{\partial y'} \right) + \epsilon \left(\frac{\partial Z'}{\partial t'} + u' \frac{\partial Z'}{\partial x'} + v' \frac{\partial Z'}{\partial y'} \right) = 0. \tag{12.34}$$

We now expand all our variables in powers of ϵ :

$$
\begin{aligned}
u' &= u_0 + \epsilon u_1 + \dots \\
v' &= v_0 + \epsilon v_1 + \dots \\
Z' &= Z_0 + \epsilon Z_1 + \dots .
\end{aligned} \tag{12.35}
$$

We next substitute (12.35) into (12.32)–(12.34) and order our equations by powers of ϵ. At zeroth order we have

$$- v_0 = -Z_{0,x'} \tag{12.36}$$

$$u_0 = -Z_{0,y'} \tag{12.37}$$

and consistent with (12.36) and (12.37)

$$\frac{\partial u_0}{\partial x'} + \frac{\partial v_0}{\partial y'} = 0. \tag{12.38}$$

Equations 12.36 and 12.37 are simply the geostrophic relations and as such they tell us nothing about time evolution. Equation 12.38 tells us that horizontal divergence is $O(\epsilon)$.

At first order in ϵ we have

$$\frac{\partial u_0}{\partial t'} + u_0 \frac{\partial u_0}{\partial x'} + v_0 \frac{\partial u_0}{\partial y'} - v_1 - \beta' y' v_0 = -Z_{1,x'} \tag{12.39}$$

$$\frac{\partial v_0}{\partial t'} + u_0 \frac{\partial v_0}{\partial x'} + v_0 \frac{\partial v_0}{\partial y'} + u_1 - \beta' y' u_0 = -Z_{1,y'} \tag{12.40}$$

$$\frac{\partial u_1}{\partial x'} + \frac{\partial v_1}{\partial y'} + \left(\frac{\partial Z_0}{\partial t'} + u_0 \frac{\partial Z_0}{\partial x'} + v_0 \frac{\partial Z_0}{\partial y'} \right) = 0. \tag{12.41}$$

We next differentiate (12.39) with respect to y', and (12.40) with respect to x', and subtract the results just as in Section 12.2 to obtain

$$\left(\frac{\partial}{\partial t'} + u_0 \frac{\partial}{\partial x'} + v_0 \frac{\partial}{\partial y'} \right) \left(\frac{\partial v_0}{\partial x'} - \frac{\partial u_0}{\partial y'} + \beta' y' \right) + \left(\frac{\partial u_1}{\partial x'} + \frac{\partial v_1}{\partial y'} \right)$$

$$+ \beta' y' \underbrace{\left(\frac{\partial v_0}{\partial y'} + \frac{\partial u_0}{\partial x'} \right)}_{=0} + \underbrace{\left(\frac{\partial u_0}{\partial x'} + \frac{\partial v_0}{\partial y'} \right)}_{=0} \left(\frac{\partial v_0}{\partial x'} - \frac{\partial u_0}{\partial y'} \right) = 0.$$

$$\tag{12.42}$$

Using (12.41) we finally get[2]

$$\frac{D}{Dt}\bigg|_0 \left(\frac{\partial v_0}{\partial x'} - \frac{\partial u_0}{\partial y'} + \beta' y' \right) - \frac{D}{Dt}\bigg|_0 Z_0 = 0. \tag{12.43}$$

Equation 12.43 is simply (12.11) where the advective velocities and the relative vorticity are evaluated geostrophically. Equations 12.36, 12.37, and 12.43 completely determine the zeroth order fields, but note that we had to go to first order in ϵ to get (12.43). Not surprisingly, the evolution of quasi-geostrophic flow is completely determined by the vorticity equation. (Note that $|f| \gg |\zeta|$.)

12.4   Quasi-geostrophy in a stratified, compressible atmosphere

Given the close relation we have noted in Chapter 11 between the shallow water equations and the equations for internal waves in a deep atmosphere, we may reasonably anticipate that the quasi-geostrophic equations for a deep atmosphere will be similar to those we have just obtained.

Our equations of motion in $\log -p$ coordinates are

$$\left(\frac{\partial}{\partial t} + u\frac{\partial}{\partial x} + v\frac{\partial}{\partial y} + w^*\frac{\partial}{\partial z^*} \right) u - fv = -\Phi_x \tag{12.44}$$

$$\left(\frac{\partial}{\partial t} + u\frac{\partial}{\partial x} + v\frac{\partial}{\partial y} + w^*\frac{\partial}{\partial z^*} \right) v + fu = -\Phi_y \tag{12.45}$$

$$\left(\frac{\partial u}{\partial x} + \frac{\partial v}{\partial y} \right) + \epsilon^{z^*}\frac{\partial}{\partial z^*}(e^{-z^*} w^*) = 0 \tag{12.46}$$

$$\frac{\partial \Phi}{\partial z^*} = RT \tag{12.47}$$

$$\left(\frac{\partial}{\partial t} + u\frac{\partial}{\partial x} + v\frac{\partial}{\partial y} \right) T + w^* \left(\frac{\partial T}{\partial z^*} + \frac{RT}{c_p} \right) = 0. \tag{12.48}$$

---

[2]The notation $\frac{D}{Dt}\big|_0$ refers to the standard substantial derivative where the advecting velocities are evaluated geostrophically.

If

$$
\begin{aligned}
w^* &= Ww' \\
u, v &= Uu', Uv' \\
z^* &= Hz' \\
x, y &= Lx', Ly'
\end{aligned}
$$

then we know from Section 12.3 that

$$
W \sim \frac{H}{L} U \epsilon
$$

because the geostrophic divergence ~ 0. Hence the vertical advections will be at least a factor ϵ smaller than the horizontal advections. However, the latter are already $O(\epsilon)$ compared to the Coriolis term. Thus *vertical advections will not enter our equations* at either zeroth or first order in ϵ, and Equations 12.44 and 12.45 are essentially *identical to our shallow water equations* for u and v. Thus at zeroth order

$$
- f_0 v_G = -\Phi_x \tag{12.49}
$$

$$
+ f_0 u_G = -\Phi_y \tag{12.50}
$$

(where $f = f_0 + \beta y$). Similarly, to first order

$$
\left(\frac{\partial}{\partial t} + u_G \frac{\partial}{\partial x} + v_G \frac{\partial}{\partial y} \right) \left(\frac{\partial v_G}{\partial x} - \frac{\partial u_G}{\partial y} + f \right) + f_0 \left(\frac{\partial u_1}{\partial x} + \frac{\partial v_1}{\partial y} \right) = 0. \tag{12.51}
$$

(N.B. we are retaining dimensional variables.)

Equation 12.46 relates w^* to

$$
\frac{\partial u_1}{\partial x} + \frac{\partial v_1}{\partial y}.
$$

Equation 12.47 allows us to rewrite (12.48) as

$$
\left(\frac{\partial}{\partial t} + u \frac{\partial}{\partial x} + v \frac{\partial}{\partial y} \right) \frac{\partial \Phi}{\partial z^*} + w^* R \left(\frac{\partial T}{\partial z^*} + \frac{RT}{c_p} \right) = 0. \tag{12.52}
$$

By analogy with our shallow water analysis we can, to lowest order, replace

$$\frac{\partial}{\partial t} + u\frac{\partial}{\partial x} + v\frac{\partial}{\partial y}$$

in (12.52) with

$$\frac{\partial}{\partial t} + u_G\frac{\partial}{\partial x} + v_G\frac{\partial}{\partial y}.$$

Also, we can replace

$$\frac{\partial T}{\partial z^*} + \frac{RT}{c_p}$$

with its horizontal average

$$\frac{\partial \bar{T}}{\partial z^*} + \frac{R\bar{T}}{c_p}$$

(Why?).

Thus we have

$$w^* = \frac{\left(\frac{\partial}{\partial t} + u_G\frac{\partial}{\partial x} + v_G\frac{\partial}{\partial y}\right)\frac{\partial \Phi}{\partial z^*}}{R\left(\frac{d\bar{T}}{dz^*} + \frac{R\bar{T}}{c_p}\right)}. \tag{12.53}$$

Equations 12.53 and 12.46 then give

$$\frac{\partial u_1}{\partial x} + \frac{\partial v_1}{\partial y} = e^{z^*}\frac{\partial}{\partial z^*}\left\{\frac{e^{-z^*}\left(\frac{\partial}{\partial t} + u_G\frac{\partial}{\partial x} + v_G\frac{\partial}{\partial y}\right)\frac{\partial \Phi}{\partial z^*}}{R\left(\frac{d\bar{T}}{dz^*} + \frac{R\bar{T}}{c_p}\right)}\right\}. \tag{12.54}$$

If we let

$$S = R\left(\frac{d\bar{T}}{dz^*} + \frac{R\bar{T}}{c_p}\right),$$

(12.54) becomes

$$\frac{\partial u_1}{\partial x} + \frac{\partial v_1}{\partial y} = \left(\frac{\partial}{\partial t} + u_G\frac{\partial}{\partial x} + v_G\frac{\partial}{\partial y}\right)\left\{e^{z^*}\frac{\partial}{\partial z^*}\left(\frac{e^{-z^*}}{S}\frac{\partial \Phi}{\partial z^*}\right)\right\}$$
$$+ \underbrace{\frac{1}{S}\left(\frac{\partial u_G}{\partial z^*}\frac{\partial}{\partial x} + \frac{\partial v_G}{\partial z^*}\frac{\partial}{\partial y}\right)\frac{\partial \Phi}{\partial z^*}}_{=0 \text{ by geostrophy}}. \tag{12.55}$$

With (12.55), (12.51) becomes

$$\left(\frac{\partial}{\partial t} + u_G\frac{\partial}{\partial x} + v_G\frac{\partial}{\partial y}\right)\left(\frac{\partial u_G}{\partial x} - \frac{\partial u_G}{\partial y} + f\right)$$

$$+ \left(\frac{\partial}{\partial t} + u_G\frac{\partial}{\partial x} + v_G\frac{\partial}{\partial y}\right)\left(e^{z^*}\frac{\partial}{\partial z^*}\left(\frac{f}{S}e^{-z^*}\frac{\partial\Phi}{\partial z^*}\right)\right) = 0$$

or

$$\left(\frac{\partial}{\partial t} + u_G\frac{\partial}{\partial x} + v_G\frac{\partial}{\partial y}\right)$$

$$\left\{\frac{\partial v_G}{\partial x} - \frac{\partial u_G}{\partial y} + f + e^{z^*}\frac{\partial}{\partial z^*}\left(\frac{f_0}{S}e^{-z^*}\frac{\partial\Phi}{\partial z^*}\right)\right\} = 0. \quad (12.56)$$

12.4.1 Pseudo-potential vorticity

The quantity in brackets in Equation 12.56 is called the *pseudo-potential vorticity* since it is conserved not on particle trajectories but on their horizontal projections. The relation between (12.56) and (12.43) is much what we would expect from our earlier comparison of the equations for shallow water waves and internal waves. Using (12.49) and (12.50), (12.56) becomes

$$\left(\frac{\partial}{\partial t} - \frac{1}{f_0}\frac{\partial\Phi}{\partial y}\frac{\partial}{\partial x} + \frac{1}{f_0}\frac{\partial\Phi}{\partial x}\frac{\partial}{\partial y}\right)\left\{\frac{1}{f_0}\left(\frac{\partial^2\Phi}{\partial x^2} + \frac{\partial^2\Phi}{\partial y^2}\right) + f\right.$$

$$\left. + e^{z^*}\frac{\partial}{\partial z^*}\left(\frac{f_0}{S}e^{-z^*}\frac{\partial\Phi}{\partial z^*}\right)\right\} = 0. \quad (12.57)$$

The quasi-geostrophic approximation was originally developed by Charney (1948). Note that the height field completely determines quasi-geostrophic motion – even its time evolution.

Exercises

12.1 Using (12.43), (12.36), and (12.37), evaluate the quasi-geostrophic potential vorticity in terms of Z using *dimensional* variables.

Identify the Rossby radius. What does it tell you about the horizontal range of influence of point sources of vorticity? (See Pedlosky, 1979, pp. 101–5.) How does this account for the maximum Rossby wave frequency you obtained in Exercise 11.3?

12.2 Use (12.57) to derive the vertical structure equation for a linearized internal Rossby wave on a basic zonal flow $u_0(z^*)$. Show that when $u_0 = $ constant we regain the results of Chapter 11.

12.3 Show that for flow rotating about a point $\vec{x} = 0$ with constant angular momentum/unit mass (i.e., $u_\phi = c/r$), vorticity is zero everywhere except at $\vec{x} = 0$. Explain this result in terms of the discussion in Section 12.1.

Chapter 13

The generation of eddies by instability, 1

Supplemental reading:

Holton (1979), sections 9.2, 9.3

Pedlosky (1979), sections 7.1–7.3

13.1 Remarks

In Chapters 8–11 we examined the wave properties of the atmosphere under a variety of circumstances. We have also considered the interactions of waves with mean flows. In Chapter 7, we noted that eddies would have to be involved in transporting heat between the tropics and the poles. Thus far our study of waves has not provided much insight into this matter. As it turns out, vertically propagating stationary Rossby waves do carry heat poleward. This heat transport, while not insignificant, is largely a byproduct of the fact that the wave momentum flux acts to reduce shears, and geostrophic adjustment involves a concomitant reduction of meridional temperature gradients. Unfortunately, we will not have time to study these mechanisms in these notes. However, the quasi-geostrophic framework established in Chapter 12 greatly simplifies such studies. This framework will be used to study how travelling disturbances arise in the atmosphere. We will also sketch (in Chapter 14) some results which suggest that these travelling disturbances play the major role in determining the global, temperature distribution. The generation of such disturbances involves hydrodynamic instability, and before diving into this problem in a meteorological context, it will be useful to examine stability in simpler situations.

Before beginning this topic we should recall possible wave (eddy) sources considered thus far:

1. Direct forcing as produced by tidal heating, flow over mountains, or flow through quasi-stationary inhomogeneities in heating.

2. Resonant free oscillations. These are presumably preferred responses to any 'noise'. However, given the presence of dissipation it is not clear what precisely maintains the free Rossby waves observed in the atmosphere. Moreover, when the relative phase speeds of Rossby waves become small compared to *variations* in the mean zonal flow, the free Rossby waves cease to exist. Observed free Rossby waves have large phase speeds.

The bulk of the *travelling* disturbances in the (lower) atmosphere and oceans are due to neither of the above, but arise as instabilities on the mean flow.

13.2   Instability

We shall use the word instability to refer to any situation where a perturbation extracts 'energy' from the unperturbed flow. The word 'energy' is surrounded by quotes because the concept of energy is not always unambiguous. Crudely, an instability grows at the expense of the basic flow. The precise sense in which this is true may have to be elaborated on.

This topic has been studied for well over a century. It is still a major area of research with many areas of uncertainty and ignorance. In these last two chapters, we can only hope to convey a taste of what is an important, interesting, and difficult subject.

The most commonly studied approach to instability is by way of what are called 'normal mode' instabilities. We earlier referred to the

solutions of the homogeneous perturbation equations as free oscillations. These were normal mode solutions in the sense that an initial
perturbation of such form would continue in that form; this would not
be true for an arbitrary (or non-normal mode) initial perturbation. In
particularly simple situations, we solved for the frequencies of these
oscillations. In the situations we studied (U_0 = constant) these frequencies were real, but in other situations these frequencies can have
an imaginary part, σ_i. When the sign of σ_i is such as to imply exponential growth in time, the basic state is said to be unstable with respect
to normal mode perturbations.

Most of our attention will be devoted to these normal mode instabilities, but you should be aware that these are not the only cases of
instability. It frequently occurs that arbitrary initial perturbations have
algebraic rather than exponential growth. In addition disturbances may
have initial algebraic growth followed by algebraic decay. Such situations have frequently been ignored in the past because of the eventual
decay, but clearly temporary growth – especially when rapid – may be
of considerable practical importance. The plethora of possibilities may
be a little confusing, but it is important to be aware of them. We will
present an example of a non-normal mode instability at the end of this
chapter.

13.2.1 Buoyant convection

This particular example is chosen as a simple example of the traditional normal mode approach to instability. The mathematical basis
for our inquiry is the treatment of simple internal gravity waves in a
Boussinesq fluid given in Chapter 8. Recall that we were looking at
two-dimensional perturbations in the x, z-plane on a static basic state
with Brunt-Vaisala frequency N. The perturbation vertical velocity, w,
satisfied the following equation:

$$w_{zz} + \left\{ \left(\frac{N^2}{\sigma^2} - 1 \right) k^2 \right\} w = 0, \tag{13.1}$$

where w had an x, t dependence of the form $e^{i(kx-\sigma t)}$. As boundary
conditions we will take

$$w = 0 \text{ at } z = 0, H. \tag{13.2}$$

Thus far we have introduced nothing new. However, we will now take $N^2 < 0$! As in our earlier analysis, (13.1) has solutions of the form

$$w = \sin \lambda z, \tag{13.3}$$

where

$$\lambda = \left(\frac{N^2}{\sigma^2} - 1\right)^{1/2} k = \frac{n\pi}{H}, \quad n = 1, 2, \dots . \tag{13.4}$$

Solving for σ^2 we again get

$$\sigma^2 = \frac{N^2}{1 + \frac{n^2\pi^2}{k^2 H^2}}, \tag{13.5}$$

but now $\sigma^2 < 0$ and

$$\sigma = \pm i \left\{ \frac{-N^2}{1 + \frac{n^2\pi^2}{k^2 H^2}} \right\}^{1/2}. \tag{13.6}$$

The largest growth rate is associated with $n = 1$ and $k = \infty$, and is given by

$$\sigma = i(-N^2)^{1/2}. \tag{13.7}$$

13.2.2 Rayleigh–Benard instability

In reality viscosity and thermal conductivity tend to suppress small scales. Very crudely, they produce a damping rate

$$d \sim \nu(\lambda^2 + k^2) \tag{13.8}$$

so that (13.6) becomes

$$\sigma \approx i \left\{ \underbrace{\left\{ \frac{-N^2}{1 + \frac{n^2\pi^2}{k^2 H^2}} \right\}^{1/2}}_{A} - \underbrace{\nu k^2 \left(1 + \frac{n^2\pi^2}{k^2 H^2}\right)}_{B} \right\}. \tag{13.9}$$

Term A exceeds term B only over a finite range of k, provided $-N^2$ is large enough. A schematic plot of A and B illustrates this. As an

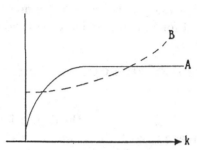

Figure 13.1: A plot of the buoyant growth term, A, and the diffusive damping term, B, in Equation 13.9.

exercise you will work out the critical value of N^2 and the optimum k. The above problem is usually referred to as *Rayleigh-Benard* instability for stress-free boundaries. In the atmosphere and oceans, even slightly unstable conditions lead to large values of $-N^2$; maximum growth rates very nearly approach the value given by (13.7) for

$$k > O(\frac{1}{H}).$$

13.2.3 Convective adjustment and gravity wave breaking

Such growth times are so much shorter than typical time scales for medium–large scale motions as to imply that convection will prevent $N^2 \lesssim 0$ over the longer time scales (at least above the surface boundary layer). This process of *convective adjustment* is of substantial importance in atmospheric and oceanic physics and in modelling.

A practical application of convective adjustment arises in connection with vertically propagating gravity waves. Recall that such waves increase in amplitude as $e^{z*/2}$ and also that such waves are approximately solutions to the nonlinear equations. Thus at some height the temperature field associated with the wave should become statically unstable – but for the onset of convection. The intensity of the convection (turbulence) ought to be proportional to the time rate of change in temperature which the wave would produce in the absence of turbulence. This mechanism is currently believed to account for the turbulence in the mesosphere called for in composition calculations.

13.2.4   Reversal of mesopause temperature gradient

The 'wave breaking' also must lead to the deposition of the waves' momentum flux in the mean flow. If the waves originate in the troposphere (and thus have small values of phase speed c), then the deposition of their momentum flux will lead to a slowing of mesospheric zonal winds and the reversal of the pole–pole temperature gradient as observed at the mesopause (Why?).

13.2.5   Kelvin-Helmholtz instability

This problem consists in the investigation of the free solutions in a stratified (constant N^2), non-rotating, infinite Boussinesq fluid with the following basic velocity profile:

$$
\begin{aligned}
u_0 &= U \text{ for } z > 0 \\
u_0 &= -U \text{ for } z < 0.
\end{aligned}
\tag{13.10}
$$

Away from $z = 0$, solutions of the form $e^{ik(x-ct)}$ satisfy

$$
\frac{d^2 w'}{dz^2} + \left[\frac{N^2}{(u_0 - c)^2} - k^2 \right] w' = 0.
\tag{13.11}
$$

For an infinite fluid our boundary conditions are that w remain bounded as $|z| \to \infty$. Also, if

$$
\frac{N^2}{(u_0 - c)^2} - k^2
$$

is positive for either $z > 0$ or $z < 0$, we require the appropriate radiation condition.

The discontinuity in u_0 at $z = 0$ means that we have different solutions for $z > 0$ and $z < 0$. At $z = 0$ we require continuity of perturbation vertical displacement and pressure. It is easy to show that

$$\frac{p'}{\rho_0} = \frac{u_0 - c}{ik}\frac{dw'}{dz} \tag{13.12}$$

and

$$w' = ik(u_0 - c)Z, \tag{13.13}$$

where Z = vertical displacement.

Thus we require continuity of

$$(u_0 - c)\frac{dw'}{dz}$$

and

$$\frac{w'}{(u_0 - c)} \tag{13.14}$$

at $z = 0$.

It is also easy to show that no free solutions exist for $|\mathrm{Re}(c)| > U$. (Why? Hint: Use Eliassen-Palm theorems.)

Now let

$$n_1^2 = \left(\frac{N^2}{(U - c)^2} - k^2\right) \tag{13.15}$$

$$n_2^2 = \left(\frac{N^2}{(U + c)^2} - k^2\right). \tag{13.16}$$

Then for $z > 0$

$$w' = A_1 e^{\pm in_1 z}, \tag{13.17}$$

and for $z < 0$

$$w' = A_2 e^{\pm in_2 z}. \tag{13.18}$$

The choice of sign in (13.17) and (13.18) is made to satisfy boundary conditions as $z \to \pm\infty$. Once these choices are made, (13.14) yields the relation between c and k. Solving for this calls for a fair amount of algebra which can be found in Lindzen (1974) and Lindzen and Rosenthal (1976). Here we shall merely cite the results.

13.2.6   Radiating and growing solutions

(i) For $k < \frac{N}{U}$ we have a solution where $c_i = c_r = 0$,

$$
\begin{aligned}
w &= A e^{i n_1 z} \text{ for } z > 0 \\
&= -A e^{i n_2 z} \text{ for } z < 0.
\end{aligned}
\tag{13.19}
$$

(ii) For $\frac{N}{2U} < k < \frac{N}{\sqrt{2}U}$ we also have solutions given by (13.19) where $c_i = 0$ and

$$
c_r = \pm \left[\frac{N^2}{2k^2} - U^2 \right]^{1/2}.
\tag{13.20}
$$

(iii) For $k > \frac{N}{\sqrt{2}U}$ we have solutions for which $c_r = 0$ and

$$
c_i = \left(U^2 - \frac{N^2}{2k^2} \right)^{1/2},
\tag{13.21}
$$

where

$$
\begin{aligned}
w &= A_1 e^{-nz} \text{ for } z > 0 \\
&= A_2 e^{n^* z} \text{ for } z < 0
\end{aligned}
\tag{13.22}
$$

and

$$
n^2 = -\frac{k^2 (U + i c_i)^2}{(U - i c_i)^2},
\tag{13.23}
$$

where n = that root of (13.23) with a positive real part. Also,

$$
A_2 = -\frac{U + i c_i}{U - i c_i} A_1.
\tag{13.24}
$$

The above results show that a strong shear zone can generate both growing interfacial disturbances confined to the shear zone and internal gravity waves propagating away from the shear zone. In each case the real phase speeds are confined between $\pm U$.

The instabilities are known as Kelvin-Helmholtz waves. Both these and the radiating gravity waves play a major role in clear air turbulence.

Before leaving this problem, consider the solution consisting in gravity waves radiating away from the shear layer a little further. When waves radiate away from a region without any waves approaching that region, we have what amounts to an infinite reflection coefficient. This is an extreme example of over-reflection. Over-reflection refers to situations where waves are reflected with reflection coefficients that exceed one. You may confirm for yourself in the present problem that gravity waves incident on the shear layer with phase speeds between $\pm U$ will, in general, be over-reflected. Finally, it should be noted that if we had a reflecting boundary below and/or above the shear layer, then over-reflection could lead to growing modes. A wave approaching the shear layer would be over-reflected and returned to the reflecting boundary with increased amplitude. Reflection at the boundary would return the wave to the over-reflecting shear layer for further amplification. Such a process could obviously lead to continuous magnification – provided that reflected and over-reflected waves were in phase so as to avoid destructive interference. This is described in detail in Lindzen and Rosenthal (1976).

13.3 Instability of meteorological disturbances; baroclinic and barotropic instability

Now that we have some idea of the formal approach to linear instability theory, we will look at the rather difficult question of whether instability can account for the travelling disturbances we saw on weather maps. Naturally, our approach will not be comprehensive, but I will attempt to deal with a few aspects which I believe to be particularly central.

Our starting point in the theoretical analysis of this problem will be the quasi-geostrophic equations of Chapter 12. We will consider a basic state consisting of a purely zonal flow $\bar{u}(y, z)$,

$$\bar{u}(y, z) = \frac{-1}{f_0} \frac{\partial \bar{\Phi}}{\partial y} \qquad (13.25)$$

$$\bar{\Phi} = \bar{\Phi}(y, z) \qquad (13.26)$$

$$\bar{v} \equiv 0.$$

If we write $\Phi = \bar{\Phi} + \Phi'$ (also $u = \bar{u} + u', v = v', w = w'$), (12.57) (or (12.56)) upon linearization becomes

$$\left(\frac{\partial}{\partial t} + \bar{u}_0 \frac{\partial}{\partial x}\right) q' + \underbrace{\frac{1}{f_0} \frac{\partial \Phi'}{\partial x}}_{v'} \bar{q}_y = 0, \tag{13.27}$$

where

$$q = \frac{1}{f_0}\left(\frac{\partial^2 \Phi}{\partial x^2} + \frac{\partial^2 \Phi}{\partial y^2}\right) + f + e^{z*}\frac{\partial}{\partial z^*}\left(\frac{f_0}{S}e^{-z*}\frac{\partial \Phi}{\partial z^*}\right) \tag{13.28}$$

$$\bar{q} = \frac{1}{f_0}\frac{\partial^2 \bar{\Phi}}{\partial y^2} + f + e^{z*}\frac{\partial}{\partial z^*}\left(\frac{f_0}{S}e^{-z*}\frac{\partial \bar{\Phi}}{\partial z^*}\right) \tag{13.29}$$

$$\frac{\partial \bar{q}}{\partial y} = -\frac{\partial^2 \bar{u}}{\partial y^2} + \beta - e^{z*}\frac{\partial}{\partial z^*}\left(\frac{f_0^2}{S}e^{-z*}\frac{\partial \bar{u}}{\partial z*}\right) \tag{13.30}$$

$$q' = \frac{1}{f_0}\left(\frac{\partial^2 \Phi'}{\partial x^2} + \frac{\partial^2 \Phi'}{\partial y^2}\right) + e^{z*}\frac{\partial}{\partial z*}\left(\frac{f_0}{S}e^{-z*}\frac{\partial \Phi'}{\partial z*}\right) \tag{13.31}$$

Traditionally, one has also made the following approximations

$$H = \text{constant} \tag{13.32}$$

$$z = Hz* \tag{13.33}$$

$$w = Hw*, \tag{13.34}$$

in which case (13.30) becomes

$$\begin{aligned}
\frac{\partial \bar{q}}{\partial y} &= -\frac{\partial^2 \bar{u}}{\partial y^2} + \beta - e^{z/H}H\frac{\partial}{\partial z}\left(\frac{f_0^2}{S}e^{-z/H}H\frac{\partial \bar{u}}{\partial z}\right) \\
&= -\frac{\partial^2 \bar{u}}{\partial y^2} + \beta - e^{z/H}\frac{\partial}{\partial z}\left(\frac{f_0^2}{N^2}e^{-z/H}\frac{\partial \bar{u}}{\partial z}\right) \\
&= \underbrace{-\frac{\partial^2 \bar{u}}{\partial y^2} + \beta + \frac{1}{H}\frac{f_0^2}{N^2}\frac{\partial \bar{u}}{\partial z} - \frac{f_0^2}{N^2}\frac{\partial^2 \bar{u}}{\partial z^2}}_{\text{assuming } N^2 \text{ is independent of } z}. \tag{13.35}
\end{aligned}$$

Note

$$\frac{H^2}{S} = \frac{H^2}{RH(\frac{\partial T_0}{\partial z} + \frac{g}{c_p})} = \frac{1}{(\frac{g}{T_0}(\frac{dT_0}{dz} + \frac{g}{c_p}))} = \frac{1}{N^2}.$$

Also,

$$q' = \frac{1}{f_0}(\nabla_H^2 \Phi') + e^{z/H}\frac{\partial}{\partial z}\left(\frac{f_0}{N^2}e^{-z/H}\frac{\partial \Phi'}{\partial z}\right). \tag{13.36}$$

Consistent with this approximation, (12.52) becomes

$$\left(\frac{\partial}{\partial t} + u_G\frac{\partial}{\partial x} + v_G\frac{\partial}{\partial y}\right)\frac{\partial \Phi}{\partial z} + wN^2 = 0, \tag{13.37}$$

which becomes, on linearization,

$$\left(\frac{\partial}{\partial t} + \bar{u}\frac{\partial}{\partial x}\right)\frac{\partial \Phi'}{\partial z} - \frac{\partial \Phi'}{\partial x}\frac{\partial \bar{u}}{\partial z} + w'N^2 = 0. \tag{13.38}$$

To summarize, the equation for Φ' that we obtain from (13.27) (using (13.36) and (13.34)) is:

$$\left(\frac{\partial}{\partial t} + \bar{u}\frac{\partial}{\partial x}\right)\left(\nabla_H^2 \Phi' + e^{z/H}\frac{\partial}{\partial z}\left(\epsilon e^{-z/H}\frac{\partial \Phi'}{\partial z}\right)\right)$$
$$+ \frac{\partial \Phi'}{\partial x}\left(\beta - \frac{\partial^2 \bar{u}}{\partial y^2} - e^{z/H}\frac{\partial}{\partial z}\left(\epsilon e^{-zh}\frac{\partial \bar{u}}{\partial z}\right)\right) = 0, \tag{13.39}$$

where

$$\epsilon \equiv f_0^2/N^2.$$

Our lower boundary condition

$$w' = 0 \text{ at z} = 0 \tag{13.40}$$

becomes (using (13.38))

$$\left(\frac{\partial}{\partial t} + \bar{u}\frac{\partial}{\partial x}\right)\frac{\partial \Phi'}{\partial z} - \frac{\partial \Phi'}{\partial x}\frac{\partial \bar{u}}{\partial z} = 0 \text{ at } z = 0. \tag{13.41}$$

As an upper boundary condition we either assume (13.41) to hold at some upper lid or require suitable boundedness (or the radiation condition) as $z \to \infty$.

Finally, we restrict ourselves to plane wave solutions of the form $e^{ik(x-ct)}$ (recall that for instability c must have a positive imaginary part) so that (13.39) and (13.41) become

$$(\bar{u} - c)\left(\frac{\partial^2 \Phi'}{\partial y^2} - k^2 \Phi' + e^{z/H}\frac{\partial}{\partial z}\left(\epsilon e^{-z/H}\frac{\partial \Phi'}{\partial z}\right)\right) + \Phi' \bar{q}_y = 0 \qquad (13.42)$$

and

$$(\bar{u} - c)\frac{\partial \Phi'}{\partial z} - \Phi'\frac{\partial \bar{u}}{\partial z} = 0 \text{ at } z = 0. \qquad (13.43)$$

Equations 13.42 and 13.43, although linear, are still very hard to solve. Indeed, with few exceptions, only numerical solutions exist. Needless to say, we shall not solve (13.42) and (13.43) here. We shall, however, establish some general properties of the solutions, and find one very simple solution.

13.3.1   A necessary condition for instability

Those readers who have studied fluid mechanics are likely to be familiar with *Rayleigh's inflection point theorem*. This theorem states that a necessary condition for the instability of plane parallel flow $\bar{u}(y)$ in a non-rotating, unstratified fluid is that $\frac{d^2\bar{u}}{dy^2}$ change sign somewhere in the fluid. Now $\frac{d^2\bar{u}}{dy^2}$ is simply $\bar{q}_y$ in such a fluid. Kuo (1949) extended this result to a rotating barotropic fluid (essentially a fluid where horizontal velocity is independent of height) and showed that $\beta - \frac{d^2\bar{u}}{dy^2}$ must change sign. A far more general result concerning $\bar{q}_y$ was obtained from (13.42) and the boundary conditions by Charney and Stern (1962). The afore-mentioned results turn out to be special cases of the more general result. We will procede to derive Charney and Stern's result.

As usual, we will assume a channel geometry where $\Phi' = 0$ at $y = y_1, y_2$. The formal derivation of our condition is quite simple. We divide (13.42) by $(\bar{u} - c)$, multiply it by $e^{-z/H}\Phi'^*$ (Φ'^* is the complex conjugate of Φ'), and integrate over the whole y, z domain:

$$I \equiv \int_0^\infty \int_{y_1}^{y_2} e^{-z/H}\Phi'^* \left\{ \frac{\partial^2 \Phi'}{\partial y^2} \quad - \quad k^2\Phi' + e^{z/H}\frac{\partial}{\partial z}\left(\epsilon e^{-z/H}\frac{\partial \Phi'}{\partial z}\right)\right.$$

$$\left. + \quad \Phi'\frac{\bar{q}_y}{\bar{u} - c}\right\} dy\,dz$$

$$= \int_0^\infty \int_{y_1}^{y_2} e^{-z/H}\left\{ \underbrace{\frac{\partial}{\partial y}\left(\Phi'^*\frac{\partial \Phi'}{\partial y}\right)}_{integrates\ to\ zero} \quad - \quad \frac{\partial \Phi'^*}{\partial y}\frac{\partial \Phi'}{\partial y} - k^2\Phi'^*\Phi'\right.$$

$$\left. + \quad \Phi'^*\Phi'\frac{\bar{q}_y}{u - c}\right\} dy\,dz$$

$$+ \int_0^\infty \int_{y_1}^{y_2}\left\{ \underbrace{\frac{\partial}{\partial z}\left(\epsilon e^{-z/H}\Phi'^*\frac{\partial \Phi'}{\partial z}\right)}_{A} \quad - \quad \epsilon e^{-z/H}\frac{\partial \Phi'^*}{\partial z}\frac{\partial \Phi'}{\partial z}\right\} dy\,dz$$

$$= 0.$$

Using (13.43), the integral of term A above can be rewritten

$$\int_0^\infty \int_{y_1}^{y_2} A\,dy\,dz = -\int_{y_1}^{y_2} \epsilon\Phi'^*\frac{\partial \Phi'}{\partial z}\,dy\bigg|_{z=0}$$

$$= -\int_{y_1}^{y_2} \epsilon\Phi'^*\Phi'\frac{\partial \bar{u}/\partial z}{\bar{u} - c}\,dy\bigg|_{z=0}.$$

We then obtain

$$I = -\int_0^\infty \int_{y_1}^{y_2} e^{-z/H} \left\{ \left| \frac{\partial \Phi'}{\partial y} \right|^2 + k^2 |\Phi'|^2 + \epsilon \left| \frac{\partial \Phi'}{\partial z} \right|^2 \right\} dy\,dz$$

$$-\int_{y_1}^{y_2} \epsilon |\Phi'|^2 \frac{\partial \bar{u}/\partial z}{\bar{u} - c} dy \bigg|_{z=0}$$

$$+\int_0^\infty \int_{y_1}^{y_2} e^{-z/H} |\Phi'|^2 \frac{f_0 \bar{q}_y}{\bar{u} - c} dy\,dz = 0.$$

$$(13.44)$$

Now the real and imaginary parts of (13.44) must each equal zero. The imaginary part arises from the last two terms when c is complex. Let

$$c = c_r + ic_i.$$

Then

$$\frac{1}{\bar{u} - c} = \frac{1}{\bar{u} - c_r - ic_i} = \frac{\bar{u} - c_r + ic_i}{|\bar{u} - c|^2}.$$

Also let

$$P = \frac{e^{-z/H} |\Phi'|^2}{|\bar{u} - c|^2}.$$

The imaginary part of (13.44) becomes

$$c_i \left\{ -\int_{y_1}^{y_2} \epsilon P \frac{\partial \bar{u}}{\partial z} dy \bigg|_{z=0} + \int_0^\infty \int_{y_1}^{y_2} P f_0 \frac{\partial \bar{q}}{\partial y} dy\,dz \right\} = 0. \qquad (13.45)$$

Next let us define

$$\tilde{q} = \bar{q} + \epsilon \frac{1}{f_0} \frac{\partial \bar{\Phi}}{\partial z} \delta(z - 0+), \qquad (13.46)$$

where δ is the Dirac delta function. Then (13.45) becomes

$$c_i \int_0^\infty \int_{y_1}^{y_2} P \frac{\partial \tilde{q}}{\partial y} dy\,dz = 0. \qquad (13.47)$$

If $c_i \neq 0$,

$$\int_0^\infty \int_{y_1}^{y_2} P \frac{\partial \tilde{q}}{\partial y} dy dz = 0. \tag{13.48}$$

But P is positive definite; therefore, there must be some surface (possibly $z = 0+$) where $\frac{\partial \tilde{q}}{\partial y}$ changes sign. When $\bar{u} = \bar{u}(y)$, this reduces to $\beta - \bar{u}_{yy}$ changing sign, and when $\beta = 0$, it reduces to Rayleigh's inflection point theorem. (Of course, it may seem fraudulent to use quasi-geostrophic equations to derive Rayleigh's theorem – but actually it's okay. Why?)

The extension of $\bar{q}$ in (13.46) is reasonable in view of the equivalence of the following situations:

(i) letting $\frac{\partial \bar{u}}{\partial z}$ at $z = 0$ be equal to $\frac{\partial \bar{u}}{\partial z}$ at $z = 0+$, and

(ii) letting $\frac{\partial \bar{u}}{\partial z} = 0$ at $z = 0$ and having a δ-function contribution to $\frac{\partial^2 \bar{u}}{\partial z^2}$ bring $\frac{\partial \bar{u}}{\partial z}$ at $z = 0+$ to $\frac{\partial \bar{u}}{\partial z} = 0$ at $z = 0$.

In the second case, $\tilde{q}$ as defined by (13.46) is actually $\bar{q}$, the basic pseudo-potential vorticity. In many cases of practical interest, $\bar{u}_z > 0$ and the curvature terms in (13.35) are relatively small above the ground, so that $\bar{q}_y > 0$ in the bulk of the atmosphere. However, the δ-function contribution to $\frac{\partial^2 \bar{u}}{\partial z^2}$ makes $\bar{q}_y < 0$ at $z = 0+$. Thus the surface at which $\bar{q}_y$ changes sign is just at the ground.

The condition we have derived is only a necessary condition for instability, but, in practice, when it is satisfied we generally do find instability. However, there are important exceptions.

13.4   The Orr mechanism

You may have already noticed a certain difference between the convective instability problem we dealt with and the two following sections dealing with problems in plane-parallel shear flow instability: namely, the underlying physics of convective instability was clear (heavy fluid on top of light fluid), while the physics underlying shear instability was obscure; it is not in the least clear, for example, why changes of potential vorticity gradient should lead to instability. This situation is only beginning to be rectified.

Fortunately, there exists an important example of a shear amplified disturbance for which the physics is relatively clear; interestingly, the disturbance is not a normal mode. We will consider a very simple situation: namely, a basic state consisting in plane parallel flow, $U(y)$, in an unbounded, incompressible, non-rotating fluid. In this case we have a streamfunction, ψ, for the velocity perturbations, where

$$u = -\frac{\partial \psi}{\partial y} \tag{13.49}$$

$$v = \frac{\partial \psi}{\partial y}. \tag{13.50}$$

Vorticity, given by

$$\xi = \nabla \times \vec{v}$$

$$= \nabla^2 \psi \tag{13.51}$$

is conserved, so

$$\frac{\partial \xi}{\partial t} + U(y)\frac{\partial \xi}{\partial x} = 0. \tag{13.52}$$

For any initial perturbation,

$$\xi(x, y, t = 0) = F(x, y), \tag{13.53}$$

(13.52) will have a solution

$$\xi(x, y, t) = F(x - U(y)t, y); \tag{13.54}$$

that is, the original perturbation vorticity is simply carried by the basic flow. The streamfunction is obtained by inverting (13.51), while the velocity perturbations are obtained from (13.49) and (13.50). The crucial point, thus far, is that it is vorticity (rather than momentum) that is conserved.

For simplicity, we will now take

$$U(y) = Sy, \tag{13.55}$$

and

$$F(x, y) = A\cos(kx). \tag{13.56}$$

Equation 13.54 becomes

$$\xi = A\cos[k(x - Syt)]$$

$$= A\{\cos(kx)\cos(kSty)$$

$$+ \sin(kx)\sin(kSty)\}. \tag{13.57}$$

The streamfunction is easily obtained from (13.57):

$$\psi = \frac{-\xi}{k^2(1 + S^2t^2)}, \tag{13.58}$$

and

$$u = -\frac{\partial \psi}{\partial y} = \frac{-ASt}{k(1 + S^2t^2)} \sin[k(x - Syt)], \tag{13.59}$$

$$v = \frac{\partial \psi}{\partial x} = \frac{-A}{k(1 + S^2t^2)} \sin[k(x - Syt)]. \tag{13.60}$$

In general, the above solution represents algebraic decay with time. What is happening is illustrated in Figure 13.2. Phase lines, which are initially vertical (in the figure), are rotated by the basic shear, and this causes the lines to move closer together. Now vorticity is conserved and vorticity is made up of derivatives of velocity. So, as the phase lines move closer together, velocity amplitudes must decrease so that the velocity derivatives will still yield the same vorticity. You might plausibly wonder why we are devoting so much attention to a mechanism which leads to decay. But, consider next a slightly more general initial perturbation:

$$F(x, y) = A\cos(kx)\cos(my). \tag{13.61}$$

Equation 13.61 may be rewritten:

$$F(x, y) = \frac{A}{2}(\cos(kx - my) + \cos(kx + my)). \tag{13.62}$$

Now the first term on the right-hand side of (13.62) corresponds to (13.57) begun at some finite positive time, and will therefore decay.

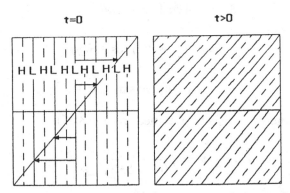

Figure 13.2: A pattern of vertical isolines of vorticity at $t = 0$ is advected by a constant shear. The dashed lines (H) correspond to high values of vorticity while the solid lines (L) correspond to low values of vorticity. At some $t > 0$, the isolines have been rotated in the direction of the shear, and the separation between the isolines has been reduced.

However, the second term corresponds to (13.57) begun at some finite negative time (i.e., its phase lines are tilted opposite to the basic shear), and this component of the initial perturbation will grow with advancing time until the phase lines are vertical. After this time, it too will decay; however, for sufficiently large m, the total perturbation velocity will grow initially (you will calculate how large m must be as an exercise), and it may grow a great deal before eventual decay sets in.

This mechanism was first discovered by Kelvin (1887); Orr (1907) suggested that this mechanism might explain why flows that are stable with respect to normal mode instabilities still become turbulent. It has been periodically rediscovered since then – but generally dismissed as less effective than exponential growth 'forever'. However, no disturbance really grows forever, and if the Orr mechanism leads to enough growth to lead to nonlinearity it will be important and the 'eventual' decay may be irrelevant. Most recently, Farrell (1987) has been arguing that the Orr mechanism is a major factor in explosive cyclogenesis, and Lindzen (1988a) has been arguing that the Orr mechanism is the underlying physical mechanism for normal mode instabilities. Unfortunately, the discussion of these tantalizingly important matters is beyond the scope of these notes.

Exercises

13.1 Using (13.9), determine the minimum value of $-N^2$ needed to produce convective instability. What is the value of k^2 for which this minimum is achieved?

13.2 Using $\sigma = i(-N^2)^{1/2}$, and $N^2 = \frac{g}{T}(\frac{\partial T}{\partial z} + \frac{g}{c_p})$, what must $\frac{\partial T}{\partial z}$ be for $\sigma = \frac{2\pi}{3\,\text{hrs}}i$? What do we learn from this?

13.3 For the Orr mechanism with the basic state and initial perturbation given by (13.55) and (13.61), respectively, find the values of m and m_0 such that for $m > m_0$ the perturbation grows before decaying. (The point is that part of (13.61) (*viz.*, (13.62)) decays and part grows; we want to know how large m must be so that the growing part grows more than the decaying part decays.)

Chapter 14

Instability 2: Energetics and climate implications

Supplemental reading:

Holton (1979), sections 9.2–9.4

Lindzen and Farrell (1980)

Charney (1973)

Lorenz (1955)

Pedlosky (1979), sections 7.7, 7.9, 7.10

14.1   Energetics of meteorological disturbances

Consistent with the approximations of Section 13.3 we may write the quasi-geostrophic equations of motions as follows:

$$\left(\frac{\partial}{\partial t} + \vec{v}_G \cdot \nabla\right)(\zeta_G + f) - f_0 e^{z/H}\frac{\partial}{\partial z}(e^{-z/H}w) = 0 \qquad (14.1)$$

$$\left(\frac{\partial}{\partial t} + \vec{v}_G \cdot \nabla\right)\Phi_z + N^2 w = 0. \qquad (14.2)$$

Presumably, these equations have an energy integral. To find this integral, let's rewrite (14.1) and (14.2):

$$\nabla^2 \psi_t + \nabla \cdot (\vec{v}_G(\nabla^2\psi + f)) - f_0 e^{z/H}\frac{\partial}{\partial z}(e^{z/H}w) = 0 \qquad (14.3)$$

$$\psi_{zt} + \nabla \cdot (\vec{v}_G\psi_z) + \frac{N^2}{f_0}w = 0, \qquad (14.4)$$

where

$$\psi \equiv \frac{\Phi}{f_0}.$$

(Recall, $\nabla \cdot \vec{v}_G = 0$.) Now multiply (14.3) by $e^{-z/H}\psi$:

$$e^{-z/H}\psi\nabla^2\psi_t \;\; + \;\; e^{-z/H}\psi\nabla \cdot (\vec{v}_G(\nabla^2\psi + f))$$

$$- \;\; f_0\psi\frac{\partial}{\partial z}(e^{-z/H}w) = 0. \qquad (14.5)$$

Now

$$\begin{aligned}
\psi\nabla^2\psi_t &= \nabla \cdot (\psi\nabla\psi_t) - \nabla\psi \cdot \nabla\psi_t \\
&= \nabla \cdot (\psi\nabla\psi_t) - \frac{1}{2}\frac{\partial}{\partial t}(|\nabla\psi|^2)
\end{aligned}$$

and

$$\begin{aligned}
\psi\nabla \cdot (\vec{v}_G(\nabla^2\psi + f)) &= \nabla \cdot (\psi\vec{v}_G(\nabla^2\psi + f)) \\
&\quad -(\nabla^2\psi + f)\underbrace{\vec{v}_G \cdot \nabla\psi}_{=0}.
\end{aligned}$$

So (14.5) becomes

$$\begin{aligned}
\frac{\partial}{\partial t}\left(e^{-z/H}\frac{|\nabla\psi|^2}{2}\right) &= e^{-z/H}\nabla \cdot (\psi\nabla\psi_t) \\
&\quad + e^{-z/H}\nabla \cdot (\psi\vec{v}_G(\nabla^2\psi + f)) \\
&\quad - f_0\psi\frac{\partial}{\partial z}(e^{-z/H}w). \qquad (14.6)
\end{aligned}$$

Now integrate (14.6) over y_1, y_2, over z between 0 and ∞, and over x (recalling periodicity in x):

$$\frac{\partial}{\partial t} \int \int \int e^{-z/H} \frac{1}{2} |\nabla \psi|^2 dx dy dz$$

$$= \underbrace{\int \int \int e^{-z/H} \nabla \cdot (\psi \nabla \psi_t) dx dy dz}_{\rightarrow 0 \ because \ \int_x \int_z e^{-z/H} \frac{\partial u_G}{\partial t} dx dz|_{y=y_1, y_2} = 0}$$

$$+ \underbrace{\int \int \int e^{-z/H} \nabla \cdot (\psi \vec{v}_G (\nabla^2 \psi + f)) dx dy dz}_{\rightarrow 0 \ because \ normal \ velocities \ at \ boundaries \ = 0}$$

$$- f_0 \int \int \int \psi \frac{\partial}{\partial z} (e^{-z/H} w) dx dy dz \qquad (14.7)$$

Clearly, the left-hand side of (14.7) is the time rate of change of kinetic energy. We turn next to (14.4). Multiply (14.4) by $e^{-z/H} \frac{f_0}{N^2} \psi_z$:

$$e^{-z/H} \frac{f_0^2}{N^2} \psi_z \psi_{zt} \quad + \quad e^{-z/H} \frac{f_0^2}{N^2} \psi_z \nabla \cdot (\vec{v}_G \psi_z)$$

$$+ \quad f_0 e^{-z/H} \psi_z w = 0. \qquad (14.8)$$

Now $\psi_z \psi_{zt} = \frac{1}{2} \frac{\partial}{\partial t} (|\psi_z|^2)$ and

$$\psi_z \nabla \cdot (\vec{v}_G \psi_z) = \frac{1}{2} \nabla \cdot (\vec{v}_G |\psi_z|^2).$$

So (14.8) becomes

$$\frac{\partial}{\partial t} \left(e^{-z/H} \frac{f_0^2}{N^2} \frac{1}{2} |\psi_z|^2 \right)$$

$$= -e^{-z/H} \frac{f_0^2}{N^2} \frac{1}{2} \nabla \cdot (\vec{v}_G |\psi_z|^2) - f_0 e^{-z/H} \psi_z w. \qquad (14.9)$$

Integrating (14.9) over x, y, z we get

$$\frac{\partial}{\partial t} \int \int \int e^{-z/H} \frac{f_0^2}{N^2} \frac{1}{2} \mid \psi_z \mid^2 dx dy dz$$

$$= - \underbrace{\int \int \int e^{-z/h} \frac{f_0^2}{N^2} \frac{1}{2} \nabla \cdot (\vec{v}_G \mid \psi_z \mid^2) dx dy dz}_{=0 \ \text{(Why?)}}$$

$$-f_0 \int \int \int \psi_z w e^{-z/H} dx dy dz. \tag{14.10}$$

The left-hand side of (14.10) is the time rate of change of something called the *available potential energy*. We will explain what this is in the next section.

Finally, adding (14.7) and (14.10) we get

$$\frac{\partial}{\partial t} \int \int \int e^{-z/H} \left\{ \frac{1}{2} |\nabla \psi|^2 + \frac{1}{2} \frac{f_0^2}{N^2} |\psi_z|^2 \right\} dx dy dz$$

$$= -f_0 \underbrace{\int \int \int \frac{\partial}{\partial z} (\psi w e^{-z/H}) dx dy dz}_{=0 \ \text{(Why?)}} .$$

The quantity
$$E = \frac{1}{2} |\nabla \psi|^2 + \frac{1}{2} \frac{f_0^2}{N^2} |\psi_z|^2 \tag{14.11}$$
is the total energy per unit mass.

Energy budgets for the atmosphere and oceans are studied at great length, but the interpretation of such studies must be approached cautiously. For example, if we divide our fields into zonally averaged parts and eddies,

$$\psi = \bar{\psi}(y, z, t) + \psi'(x, y, z, t)$$
$$T = \bar{T} + T'$$
$$w = \bar{w} + w'$$
$$\zeta = \bar{\zeta} + \zeta',$$

where $(\bar{\ }) \equiv$ zonal average of ( ), we can obtain, after modest amounts of algebra,[1]

$$\frac{\partial}{\partial t} \int_{y_1}^{y_2} \int_0^\infty e^{-z/H} \overline{\left\{ \frac{1}{2}(\nabla \psi')^2 + \frac{1}{2}\frac{f_0^2}{N^2}\left(\frac{\partial \psi'}{\partial z}\right)^2 \right\}} dydz$$

$$= \int_{y_1}^{y_2} \int_0^\infty e^{-z/H} \overline{\psi'_x \psi'_y} \; \overline{u}_y \, dydz - \int_{y_1}^{y_2} \int_0^\infty \frac{f_0^2}{N^2} \overline{\psi'_x \psi'_z} \; \overline{\psi}_{zy} e^{-z/H} dydz$$

$$= - \int_{y_1}^{y_2} \int_0^\infty e^{-z/H} \overline{u'v'} \; \overline{u}_y \, dydz$$

$$- \int_{y_1}^{y_2} \int_0^\infty \frac{g/T_0}{(<\frac{\partial T}{\partial z}> + \frac{g}{c_p})} \overline{v'T'} \; \overline{T}_y e^{-z/H} \, dydz, \qquad (14.12)$$

where T_0 = average of T over whole fluid, and $< T >$ = horizontal average of T.

It can also be shown that

$$\frac{\partial}{\partial t} \int_{y_1}^{y_2} \int_0^\infty e^{-z/H} \left\{ \frac{1}{2}(\nabla \overline{\psi})^2 + \frac{1}{2}\frac{f_0^2}{N^2}(\overline{\psi}_z)^2 \right\} dydz$$

$$= \int_{y_1}^{y_2} \int_0^\infty e^{-z/H} \overline{u'v'} \; \overline{u}_y \, dydz$$

$$+ \int_{y_1}^{y_2} \int_0^\infty \frac{g/T_0}{(<\frac{\partial T}{\partial z}> + \frac{g}{c_p})} \overline{v'T'} \; \overline{T}_y e^{-z/H} \, dydz. \qquad (14.13)$$

Equations 14.12 and 14.13, together, seem to tell us that the sum of the energy in the zonally average flow and the eddies is conserved, and that the growth of eddies occurs at the expense of the mean flow through the

---

[1]The procedure consists in averaging Equations 14.1 and 14.2 with respect to x, subtracting these averages from (14.1) and (14.2), multiplying the resulting equations by $-e^{-z/H}\psi'$ and $(f_0^2/N^2)e^{-z/H}\psi'_z$, respectively, adding the two resulting equations, and integrating over the fluid volume. The reader should go through the derivation of both Equations 14.12 and 14.13.

action of horizontal eddy stresses on $\bar{u}_y$ and horizontal eddy heat fluxes on $\bar{T}_y$. Although these remarks are subject to interpretation (and we will discuss them later) they do establish the criteria for eddy growth to be energetically consistent; in particular they state that the energy of the zonally averaged flow can only be tapped if $\bar{u}_y \neq 0$ and/or $\bar{T}_y \neq 0$ (in the above quasi-Boussinesq system).

It is, however, sometimes concluded that instabilities are caused by $\bar{u}_y$ and/or $\bar{T}_y$. This is not, in general, true. To be sure, if $\bar{u}_y = \bar{T}_y = 0$ there can be no unstable eddies. In such a case, the necessary condition for instablility given in Subsection 13.3.1 would not be satisfied. However, this condition can also fail to be satisfied when $\bar{u}_y \neq 0$ and/or $\bar{T}_y \neq 0$.

Equation 14.12 does tell us that a growing eddy will act to reduce $\bar{T}_y$ and $\bar{u}_y$. The first term on the right-hand side of (14.12) is called a barotropic conversion while the second term is called a baroclinic conversion. Instabilities of flows where $\bar{u}_y = 0$ are called *baroclinic instablilities*, while instabilities where $\bar{T}_y = 0$ (and $\bar{u}_z = 0$) are called *barotropic instabilities*. However, as we saw in Subsection 13.3.1 both instabilities depend on the meridional gradient of potential vorticity.

14.2 Available potential energy

The quantity available potential energy is at first sight a bit strange. We expect the conservation of kinetic + potential + internal energy. Now the last two items can be written as

$$P = \int \int \int gz\, \rho\, dzdxdy \tag{14.14}$$

and

$$I = \int \int \int c_v T\, \rho\, dzdxdy. \tag{14.15}$$

The former can be rewritten

$$P = \int \int \int gz\, \rho\, dzdxdy \;=\; \int \int \int_0^{p_0} z\, dp = \int \int \int_0^\infty p\, dz$$

$$= \int \int \int RT\rho\, dz = \frac{R}{c_v} I, \tag{14.16}$$

and

$$I + P = \text{total potential energy} = TPE$$
$$= (1 + \frac{c_v}{R}) \int\int\int RT\, \rho\, dzdxdy$$
$$= c_p \int\int\int T\, \rho\, dzdxdy$$
$$= \frac{5}{2} \int\int\int c^2\, \rho\, dzdxdy, \qquad (14.17)$$

where c = the sound speed. Clearly $TPE \gg K$. Noticing this disparity, Margules (1903) pointed out that without deviations from horizontal stratification no TPE is *available* to K (assuming static *stability*). Lorenz (1955) showed that what we have called available potential energy (APE) is that portion of TPE which is available to K.

Lorenz's analysis ran roughly as follows (ignoring horizontal integrals)[2]:

$$TPE = \frac{c_p}{g} \int_0^{p_0} T dp = (1 + \kappa)^{-1} \frac{c_p}{g} p_0^{-\kappa} \int_0^{\infty} p^{1+\kappa} d\Theta,$$

where $T = \Theta\, p^\kappa p_0^{-\kappa}$ and Θ is the *potential temperature*.

He then noted that the minimum TPE which motions would produce would be achieved if $p = < p >$ everywhere ($<>$ = horizontal average); that is,

$$TPE_{min} = (1 + \kappa)^{-1} \frac{c_p}{g} p_0^{-\kappa} \int_0^{\infty} < p >^{1+\kappa} d\Theta.$$

Now let

$$< APE > = < TPE - TPE_{min} >$$

$$= (1 + \kappa)^{-1} \frac{c_p}{g} p_0^{-\kappa} \int_0^{\infty} (< p^{1+\kappa} > - < p >^{1+\kappa}) d\Theta$$

---

[2] We are not particularly concerned with details since we are merely seeking an interpretation of APE.

and let

$$p = <p> + \tilde{p}$$

$$p^{1+\kappa} = <p>^{1+\kappa} \left\{ 1 + (1+\kappa)\frac{\tilde{p}}{<p>} + \frac{\kappa(1+\kappa)}{2!}\frac{\tilde{p}^2}{<p>^2} + \cdots \right\}$$

and

$$<APE> = \frac{1}{2}\kappa\frac{c_p}{g}p_0^{-\kappa}\int_0^\infty <p>^{1+\kappa}\left\langle \frac{\tilde{p}^2}{<p>^2} \right\rangle d\Theta \qquad (14.18)$$

(i.e., $<APE>$ depends on the variance of pressure over isentropic surfaces). If one wishes to deal with isobaric rather than isentropic surfaces we may use

$$\tilde{p} \approx \frac{\partial p}{\partial \theta}\tilde{\theta}$$

(where $\tilde{\theta}$ is the deviation of θ from $<\theta>$ on isobaric surfaces) to rewrite (14.18) as

$$<APE>$$
$$\approx \frac{1}{2}\kappa\frac{c_p}{g}p_0^{-\kappa}\int_0^{p_0} <p>^{-(1-\kappa)}<\theta>^2\left\{ -\frac{\partial <\theta>}{\partial p} \right\}^{-1}\left\langle \frac{\tilde{\theta}^2}{<\theta>^2} \right\rangle dp.$$

Now

$$\frac{\partial \theta}{\partial p} = -\kappa\frac{c_p}{g}\left(\frac{\partial T}{\partial z} + \frac{g}{c_p} \right).$$

Also

$$\frac{\tilde{\theta}}{\theta} \approx \frac{\tilde{T}}{T}$$

on isobaric surfaces. So

$$<APE> = \frac{1}{2}\int_0^{p_0} <T>\frac{1}{(\langle\frac{\partial t}{\partial z}\rangle + \frac{g}{c_p})}\left\langle \frac{\tilde{T}^2}{<T>^2} \right\rangle dp$$

$$= \frac{1}{2}\int_0^\infty \rho\frac{g}{<T>}\frac{1}{(\langle\frac{\partial t}{\partial z}\rangle + \frac{g}{c_p})}\langle\tilde{T}^2\rangle dz,$$

which is what we found in Section 14.1.

14.3   Some things about energy to think about

For many people there is something comforting about energetics. Yet as we have seen, it is at best a tool to establish consistency rather than causality. The problems go further than this. For example, energy is not Galilean invariant. Clearly, the kinetic energy depends on the frame of reference in which we are measuring velocity. In addition, the quantity in (14.13) whose time derivative is being taken is not the full eddy contribution to energy. You can easily prove this for yourself. Take (14.11) (which is the total energy), and substitute $\bar{\psi} + \psi'$ for ψ. The eddy contributions include cross terms which are linear in ψ' as well as the quadratic terms in (14.13). To be sure, these linear terms go to zero when averaged over x, but they are nonetheless part of the eddy energy, and their presence in some problems can lead to eddy energy not being positive-definite. This, in itself, is not as disturbing as it might appear at first sight. For example, when an eddy with a phase speed smaller than the mean flow is absorbed and reduces the mean flow speed, shouldn't we think of the eddy before it was absorbed as having negative energy insofar as its absorption led to the reduction of the mean kinetic energy? It should also be clear that this whole process would depend on the frame of reference in which speed was measured. What then is the quadratic quantity in (14.13)? Well, it is at least a pretty good measure of the overall magnitude of the eddies.

It is likely that the above is almost certain to be more confusing than edifying for many readers. However, it was merely meant to give you something to think about. In doing so you might actually avoid some confusion later on.

14.4   Two-level baroclinic model

One of the simplest models of baroclinic instability is the two-level model for the instability of a basic state $\bar{u}(z)$ in a Boussinesq fluid (where $\rho = $ constant, but $N^2 \neq 0$). For such a fluid, (14.1) and (14.2) become

$$\left(\frac{\partial}{\partial t} + u_G\frac{\partial}{\partial x} + v_G\frac{\partial}{\partial y}\right)\left(\frac{\partial v_G}{\partial x} - \frac{\partial u_G}{\partial y} + f\right) - f_0\frac{\partial w}{\partial z} = 0 \qquad (14.19)$$

$$\left(\frac{\partial}{\partial t} + u_G\frac{\partial}{\partial x} + v_G\frac{\partial}{\partial y}\right)\frac{\partial \Phi}{\partial z} + wN^2 = 0. \qquad (14.20)$$

We next seek linearized perturbations, about $u = \bar{u}(z)$, of the form $e^{ik(x-ct)}$:

$$ik(\bar{u} - c)\frac{\partial^2\Phi'}{\partial x^2} + \beta\frac{\partial\Phi'}{\partial x} = f_0^2\frac{\partial w'}{\partial z} \qquad (14.21)$$

$$ik(\bar{u} - c)\frac{\partial\Phi'}{\partial z} + \underbrace{\frac{1}{f_0}\frac{\partial\Phi'}{\partial x}\frac{\partial}{\partial z}\frac{\partial\bar{\Phi}}{\partial y}}_{\frac{\partial\Phi'}{\partial x}\frac{d\bar{u}}{dz}} + w'N^2 = 0, \qquad (14.22)$$

from which we obtain

$$ik\{-k^2(\bar{u} - c) + \beta\}\Phi' = f_0^2\frac{\partial w'}{\partial z}, \qquad (14.23)$$

$$ik\left\{(\bar{u} - c)\frac{\partial\Phi'}{\partial z} - \Phi\frac{d\bar{u}}{dz}\right\} = -N^2w'. \qquad (14.24)$$

The solution of (14.23) and (14.24), even for $\bar{u}_z = $ constant, is quite difficult. However, Phillips (1954) introduced an exceedingly crude difference approximation for which easy solutions can be obtained.

Phillips considered a fluid of depth H where $w' = 0$ at $z = 0$ and $z = H$. The vertical domain is discretized by five levels (from which we get the name two-level model):

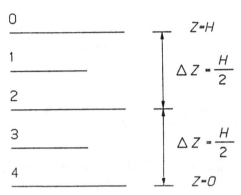

$\bar{u}$ and Φ' are evaluated at levels 1 and 3, while w' is evaluated at level 2 ($w' = 0$ at levels 0 and 4). Applying (14.23) at level 1 we get

$$ik\{-k^2(\bar{u}_1 - c) + \beta\}\Phi_1 = f_0^2\frac{w_0 - w_2}{\Delta z} = -f_0^2\frac{w_2}{\Delta z}. \qquad (14.25)$$

Applying (14.24) at level 2 we get

$$ik\left\{\left(\frac{\bar{u}_1 + \bar{u}_3}{2} - c\right)\frac{\Phi_1 - \Phi_3}{\Delta z} - \left(\frac{\Phi_1 + \Phi_3}{2}\right)\left(\frac{\bar{u}_1 - \bar{u}_3}{\Delta z}\right)\right\} = -N^2 w_2,$$

or

$$ik\{(\bar{u}_3 - c)\Phi_1 - (\bar{u}_1 - c)\Phi_3\} = -N^2\Delta z\, w_2, \qquad (14.26)$$

and evaluating (14.23) at level 3 we get

$$ik\{-k^2(\bar{u}_3 - c) + \beta\}\Phi_3 = f_0^2\frac{w_2 - w_4}{\Delta z} = f_0^2\frac{w_2}{\Delta z}. \qquad (14.27)$$

We may now use (14.25) and (14.27) to reduce (14.26) to a single equation in w_2:

$$ik\left\{(\bar{u}_3 - c)\frac{-f_0^2 w_2/\Delta z}{ik\{-k^2(\bar{u}_1 - c) + \beta\}}\right.$$
$$\left. -(\bar{u}_1 - c)\frac{f_0^2 w_2/\Delta z}{ik\{-k^2(\bar{u}_3 - c) + \beta\}}\right\} = -N^2\Delta z w_2,$$

or

$$\{(\bar{u}_3 - c)(-k^2(\bar{u}_3 - c) + \beta) + (\bar{u}_1 - c)(-k^2(\bar{u}_1 - c) + \beta)\}$$

$$= \lambda^2(-k^2(\bar{u}_1 - c) + \beta)(-k^2(\bar{u}_3 - c) + \beta), \tag{14.28}$$

where

$$\lambda^2 \equiv N^2 \frac{\Delta^2 z}{f_0^2}.$$

(It should be a matter of some concern that the fundamental horizontal scale length in this problem, λ, depends on the vertical interval, Δz, which is simply a property of our numerical procedure.) Equation 14.28 is a quadratic equation in c, which after some manipulation, may be rewritten

$$c^2\{2k^2 + \lambda^2 k^4\} \quad + \quad c\{-(4k^2 + 2\lambda^2 k^4)u_M + 2\beta(1 + k^2\lambda^2)\}$$

$$+ \quad \{u_M^2(2k^2 + \lambda^2 k^4) + u_T^2(2k^2 - \lambda^2 k^4)$$

$$- \quad 2u_M\beta(1 + k^2\lambda^2) + \beta^2\lambda^2\} = 0, \tag{14.29}$$

where

$$u_M = \frac{\bar{u}_1 + \bar{u}_3}{2} \tag{14.30}$$

$$u_T = \frac{\bar{u}_1 - \bar{u}_3}{2}. \tag{14.31}$$

Solving (14.29) for c we get

$$c = u_M - \frac{\beta(1 + k^2\lambda^2)}{(2k^2 + \lambda^2 k^4)}$$

$$\pm \underbrace{\left\{ \frac{\beta^2}{k^4(2 + \lambda^2 k^2)^2} - u_T^2 \frac{(2 - \lambda^2 k^2)}{(2 + \lambda^2 k^2)} \right\}^{1/2}}_{\equiv \delta}. \tag{14.32}$$

We will have instability when $\delta < 0$; that is, when

$$u_T^2 > \frac{\beta^2}{k^4(4 - \lambda^4 k^4)}. \tag{14.33}$$

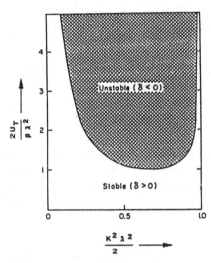

Figure 14.1: Neutral stability curve for the two-level baroclinic model.

The instability diagram for this problem is shown in Figure 14.1. The minimum value of u_T^2 needed for instability is

$$u_{T\,min}^2 = \frac{\beta^2 \lambda^4}{4}. \tag{14.34}$$

Equation 14.34 suggests that a minimum shear is needed for instability whereas Subsection 13.3.1 suggested that instability could exist for any finite shear, however small. Actually (14.34) is consistent with our earlier result. Note that for (14.19) and (14.20),

$$\bar{q}_y = \beta - \frac{f_0^2}{N^2} \frac{d^2 \bar{u}}{dz^2}. \tag{14.35}$$

Now, until recently, it was assumed that the basic flow in the 'two-level' model corresponded to a constant shear flow characterized by the estimated shear at level 2: $\frac{\bar{u}_1 - \bar{u}_3}{\Delta z}$. However, a closer study of the 'two-level' model shows that its basic flow has this shear only at level 2; at levels 0 and 4 the relevant basic shear is zero. Thus if we evaluate

(14.35) over the upper layer, we get

$$\bar{q}_y = \beta - \frac{f_0^2}{N^2}\left(\frac{0 - \frac{2u_T}{\Delta z}}{\Delta z}\right)$$

$$= \beta + \frac{2f_0^2}{N^2}\frac{u_T}{(\Delta z)^2} > 0 \quad \text{for } u_T > 0.$$

In the lower layer

$$\bar{q}_y = \beta - \frac{f_0^2}{N^2}\left(\frac{\frac{2u_T}{\Delta z} - 0}{\Delta z}\right)$$

$$= \beta - \frac{2f_0^2 u_T}{N^2(\Delta z)^2}.$$

In order for $\bar{q}_y$ to change sign we must have

$$u_T > \frac{\beta N^2(\Delta z)^2}{2f_0^2} = \frac{\beta \lambda^2}{2},$$

or

$$u_T^2 > \frac{\beta^2 \lambda^4}{4},$$

which is precisely what (14.34) says. The difference between the two-level model and a continuous fluid stems from the fact that $\bar{q}_y$ must be negative over a layer of thickness Δz in order to obtain instability in the two-level model; in a continuous fluid it suffices for $\bar{q}_y$ to be negative over an arbitrarily thin layer near the surface.

14.5 Baroclinic instability and climate

It was already suggested by Phillips that the atmosphere might be trying to achieve baroclinic neutrality, and that this would determine the meridional temperature distribution. In terms of a two-level model one gets

$$\frac{\partial \bar{T}}{\partial y} = -\frac{f_0 T_0}{g}\frac{\partial \bar{u}}{\partial z} = -\frac{f_0 T_0}{g}\frac{\beta \lambda^2}{\Delta z}$$

$$= -\frac{f_0 T_0}{g}\frac{\beta}{\Delta z}\frac{(\Delta z)^2 N^2}{f_0^2} = -\frac{f_0 T_0}{g}\frac{\beta}{\Delta z}\frac{(\Delta z)^2 \frac{g}{T_0}(\frac{\partial \bar{T}}{\partial z} + \frac{g}{c_p})}{f_0^2}$$

$$= \frac{\beta \Delta z}{f_0}\left(\frac{\partial \bar{T}}{\partial z} + \frac{g}{c_p}\right). \tag{14.36}$$

If we take (14.36) to be locally true at each latitude then

$$\frac{1}{a}\frac{\partial \bar{T}}{\partial \phi} = -\frac{\frac{2\Omega \cos\phi}{a}\Delta z}{2\Omega \sin\phi}\left(\frac{\partial \bar{T}}{\partial z} + \frac{g}{c_p}\right)$$

or (with obvious cancellation)

$$\frac{\partial \bar{T}}{\partial \phi} = -\frac{\cos\phi \,\Delta z}{\sin\phi}\left(\frac{\partial \bar{T}}{\partial z} + \frac{g}{c_p}\right). \tag{14.37}$$

If we take $\Delta z \approx 5$ km and $\frac{\partial \bar{T}}{\partial z} \approx -6.5°$/km, then (14.37) fairly uniformly underestimates the observed $\frac{\partial \bar{T}}{\partial \phi}$ at the surface by about a factor of 2 for $\phi \gtrsim 20°$. This discrepancy becomes worse when one recalls that (14.37) applies to level 2 and that the average of $\frac{\partial T}{\partial \phi}$ over the whole domain will be less than (14.37). On the other hand, (14.37) overestimates $\frac{\partial \bar{T}}{\partial \phi}$ as one approaches the equator. (Indeed (14.37) blows up at $\phi = 0$.)

Despite these problems, there are a number of reasons not to become discouraged with the suggestion that baroclinic neutrality may be relevant to climate:

(i)   Recent numerical experiments with a nonlinear two-level model strongly suggest that when $\frac{\partial T}{\partial \phi}$ for radiative equilibrium exceeds $\frac{\partial T}{\partial \phi}$ for baroclinic neutrality the system approaches baroclinic neutrality.

(ii) In the tropics we expect $\frac{\partial T}{\partial \phi}$ to be determined by the Hadley circulation – not by baroclinic instability (*viz.* (7.35)).

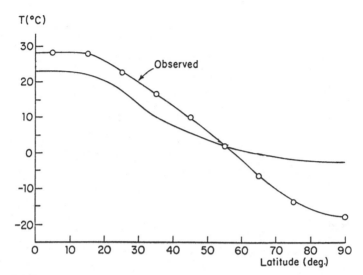

Figure 14.2: Temperature *vs.* latitude. Both the observed distribution and the result of a Hadley-baroclinic adjustment assuming constant N^2 are shown.

(iii) *A priori* we do not expect the two-level model to be quantitatively appropriate to a continuous atmosphere.

A study by Lindzen and Farrell (1980) led to the following conclusion (roughly stated): For a continuous atmosphere with radiative forcing confined below some tropopause at $z = z_B$, the appropriate baroclinically neutral profile is one where

$$\bar{q}_y = \beta - e^{z/H} \frac{\partial}{\partial z} \left(\frac{f_0^2}{N^2} e^{-z/H} \frac{\partial \bar{u}}{\partial z} \right) = 0 \text{ below } z_B \qquad (14.38)$$

and where

$$\bar{u}_z = 0 \text{ at } z = 0$$

(*viz.* (14.34)). The solution of (14.38) leads to a distribution of $\frac{\partial \bar{u}}{\partial z}$ at each latitude which in turn leads to a distribution of $\frac{\partial T}{\partial \phi}$ via the thermal wind relation. The *baroclinically adjusted* $\left(\frac{\partial T}{\partial \phi} \right)_B$ is taken to be the density weighted average of the distribution of $\frac{\partial T}{\partial \phi}$ between $z = 0$ and

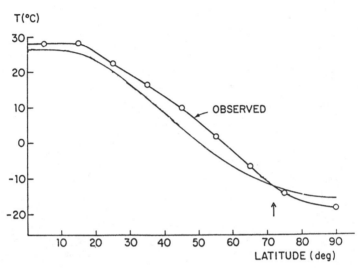

Figure 14.3: Temperature *vs.* latitude. Both the observed distribution and the result of a Hadley-baroclinic adjustment allowing for enhanced N^2 near the surface at high latitudes are shown.

$z = z_B$. We may anticipate that $\left(\frac{\partial T}{\partial \phi}\right)_B$ will come close to observations for $\phi \gtrsim 20°$. Crudely stated, Lindzen and Farrell then took the Hadley-baroclinically adjusted $\left(\frac{\partial T}{\partial \phi}\right)_{B-H}$ to be the smaller of $\left(\frac{\partial T}{\partial \phi}\right)_{Hadley}$ and $\left(\frac{\partial T}{\partial \phi}\right)_{Baroclinic}$. The resulting global distribution of $\frac{\partial T}{\partial \phi}$ can be integrated to yield $T(\phi)$ within an integration constant. This constant is chosen so that outgoing radiation integrated over the globe equals incoming radiation integrated over the globe. Results are shown in Figure 14.2; the Hadley-baroclinic temperature distribution still implies too great a heat flux. Lindzen and Farrell then noted that the assumption that N^2 was constant was not correct. Within a few kilometers of the surface, the atmosphere is substantially more stable than average at high latitudes – especially over ice-covered surfaces. This feature is readily incorporated[3]. Doing this, Lindzen and Farrell obtained the adjusted

---

[3]Of course one would like a model to predict stability – but the present approach has at least a measure of consistency.

temperatures shown in Figure 14.3. The agreement is remarkably good and strongly suggests that the observed $T(\phi)$ is largely determined by Hadley convection and baroclinic neutralization. This result is somewhat surprising insofar as current observations suggest that oceanic currents, stationary waves and transient eddies all contribute comparably to the equator–pole heat flux. However, our results suggest that baroclinic instability acts as a temperature regulator for the whole system – contributing only what is needed, in addition to the contributions of other processes to the heat flux, to achieve 'baroclinic neutrality'. This suggests that the climate is largely determined by processes which mix potential vorticity – among which processes baroclinic instability is a major contributor. These ideas are still in a rather preliminary and controversial stage, but support for them exists in numerical experiments with general circulation models (Manabe and Strickler, 1965; Manabe and Terpstra, 1974; and others) which show that the total equator–pole heat flux is not particularly sensitive to the inclusion of mountains, hydrology, and so forth – though the makeup of the heat flux, naturally enough, is. At the heart of the above problem is the nonlinear evolution of the instabilities we have touched on in the past two chapters.

Exercises

14.1 **(a)** Calculate $\bar{u}(z)$ corresponding to two-level neutrality.

 (b) Why does neutrality occur when $\bar{q}_y = 0$ in the bottom layer?

 (c) When a two-level model is unstable, how does the maximum c_i relate to the range of variation of $\bar{u}$?

 (d) How is baroclinic instability affected when N^2 is different in the two layers?

14.2 There are two classical versions of the baroclinic instability problem: the Charney (1947) problem and the Eady (1949) problem. There is also the pedagogically popular two-level problem considered in this chapter. Actually, the Eady problem is about as easy as the two-level problem. The Eady problem deals with baroclinic instability of a boussinesq fluid on an f-plane (i.e., $\beta = 0$), where the basic state consists in a constant shear, and where the fluid has top and bottom boundaries. In such a fluid, $\bar{q}_y = 0$ in the interior, which greatly simplifies the equations. Investigate the stability properties of the the Eady model. What is the physical nature of the neutral solutions?

14.3 It is clear from Section 13.3 that there need not be much mathematical difference between barotropic and baroclinic instability problems. Develop basic states for barotropic models which lead to stability problems which are mathematically identical to (a) two-level baroclinic models and (b) the Eady model.

Postscript

These notes end (as do many courses) rather abruptly. I hope to leave the reader with the sense that he or she has learned a lot. But, I would hardly wish to disguise the fact that we have barely begun the exploration of atmospheric dynamics. The nonlinear evolution and possible equilibration of instabilities which should give us the wave and flux magnitudes has only been touched on – both in these notes and in current research. A major current approach to questions of the general circulation – namely, the use of large numerical computer simulations – has not even been discussed. Areas whose impact on large-scale dynamics is almost certainly major – like boundary layer turbulence and convective cloud activity – have likewise been only peripherally dealt with in these notes. Although we have come quite far in improving our understanding of many of the phenomena and features described in Chapter 5, we are still far from a satisfactory state, and, as we have earlier noted, there exists a world of important and challenging phenomema whose scales are smaller than those discussed in Chapter 5: hurricanes, fronts, thunderstorms, squalls, to name a few. Even those topics that we have dealt with in some detail have hardly been dealt with in any measure of completeness. Whole books (in most cases several) and countless articles have been devoted specifically to instability, wave theory, the general circulation, and even tides.

The shear scope of problems which fall under the general rubric of atmospheric dynamics is so great as to lead, unfortunately, but inevitably, to a high degree of specialization. Certainly the individual problems are great enough to warrant focussed scrutiny – yet I hope that these notes have shown in a modest way how one problem often has important implications for other problems. Familiarity with topics beyond one's immediate interests is almost always helpful.

Finally, it must be clear by now that the likelihood that a reasonable number of graduate courses will serve to thoroughly cover atmospheric dynamics is highly unlikely. In this field, as in most others, learning must eventually transcend what is taught.

Appendix

Gravity wave program

A.1 Introduction

Computer algorithms are invaluable in studying and learning about waves. Rather than get involved in machine-dependent specifics, I will, in this appendix, describe the essential features of a program that I have made use of when teaching at M.I.T.

This program solves the forced gravity wave vertical structure problem, in $\log -p$ coordinates. The basic profiles of wind and temperature are specified by the user. Specific approaches towards specifying these inputs are described at the end of this appendix. As output, the program yields the finite difference solution, the WKB solution, and wave fluxes computed from the finite difference solution. The WKB solution is described in Chapter 10. Possible uses of this simple model include:

- study of the dependence of the solution on forcing and boundary conditions;

- assessing of the influence of the zonal wind and temperature profiles on the wave propagation;

- assessing the validity of the WKB solution;

- verification of the Eliassen-Palm theorems, the non-interaction conditions, and the effects of damping and forcing.

A.2   Model description

We use $\log -p$ vertical coordinates:

$$z^* = -\log \frac{p}{p_s},\tag{A.1}$$

where $p_s = 1000$ mb is a reference surface pressure. The scaled $\log -p$ coordinate

$$z_s^* = -7.4 \text{ km} \log(p/p_s)$$

$$z_s^* \approx \frac{12 \text{ km}}{\log(200\text{mb}/1000\text{mb})} \log \frac{p}{p_s}\tag{A.2}$$

is used for input/output of parameters. The model is the same considered in Chapter 10. Namely, we ignore rotation and assume that the basic state (U_0, T_0) depends only on z^* and consider perturbations for which $v' = 0$ (Equations 10.1-4). For wave-like solutions of the form $\exp(ik(x - ct))$, the perturbation equations can be reduced to a single equation (Equation 10.8):

$$\tilde{w}_{z^*z^*} + Q^2(z^*)\tilde{w} = F(z^*),\tag{A.3}$$

where $\tilde{w} = e^{-z^*/2}w^{*'}$, $w^{*'}$ being the perturbation vertical velocity in $\log -p$ coordinates. The complex index of refraction $Q(z^*)$ is

$$Q^2(z^*) = \frac{R(T_{0z^*} + \kappa T_0)}{(c - U_0)^2} + \frac{U_{0z^*z^*} + U_{0z^*}}{c - U_0} - \frac{1}{4}\tag{A.4}$$

and the right-hand-side forcing is taken as a modified Gaussian distribution modulated by $e^{-z^*/2}$, namely,

$$\begin{aligned}
F(z^*) &= e^{-z^*/2}A\{e^{-(\frac{z^*-z_f}{z_w})^2} - e^{-4}\} \\
&\qquad \text{for } |z^* - z_f| < 2z_w \\
&= 0 \\
&\qquad \text{for } |z^* - z_f| \geq 2z_w,
\end{aligned}$$

$$\tag{A.5}$$

with A, z_f, z_w being, respectively, the amplitude, centre and half-width of the forcing. A linear damping parameter a has been included in the definition of c, $\mathrm{Im}(c) = a/k$. At the ground the system is forced by a wave maker

$$\tilde{w}(z^* = 0) = \tilde{w}_{bot}. \tag{A.6}$$

At the 'top' we may either impose an artificial lid or apply the radiation condition. To impose a lid, we simply require

$$\tilde{w}(z^*_{top}) = 0. \tag{A.7}$$

The imposition of the radiation condition is more complicated. For simplicity, we will assume that at the top of our domain, Q is independent of z^*; we will also assume that forcing is essentially zero at the 'top'[1]. The solution at the top will then be given by

$$\tilde{w}(z^*_{top}) \propto e^{-iQ(z^*_{top})z^*}, \tag{A.8}$$

where Q is chosen as that square root of Equation A.4 for which $\mathrm{Im}(Q) < 0$. Equation A.8 is equivalent to setting

$$\frac{d\tilde{w}}{dz^*} = -iQ\tilde{w} \text{ at } z^* = z^*_{top}. \tag{A.9}$$

In addition to wave fields, it is also useful to calculate the following wave fluxes:

$$F_m = e^{-z^*}\overline{u'w^{*'}} \tag{A.10}$$

$$F_e = e^{-z^*}\overline{\Phi'w^{*'}}, \tag{A.11}$$

where the bar denotes zonal average,

$$\overline{ab} = \frac{1}{2}\mathrm{Re}(ab^\dagger), \tag{A.12}$$

with $b^\dagger$ denoting the complex conjugate of b. Φ' and $w^{*'}$ are computed from $w^{*'}$ using Equations 10.5 and 10.6. The wave fluxes in $\log -p$ coordinates are similar to those in z coordinates: here e^{-z^*} plays the role of density (*viz.* Exercise 8.1).

---

[1]In view of our choice of forcing, this requires that our top be higher than $z_f + 2z_w$.

A.3 Numerics

The grid is specified as follows:

$$z_k = k\Delta, k = 1, \ldots, K, \tag{A.13}$$

where the mesh size Δ is given by

$$\Delta = \frac{z^*_{top}}{K+1}. \tag{A.14}$$

Notice that neither the ground nor the 'top' of the model is included as a grid point. We approximate the second z^*-derivative by the standard formula

$$\tilde{w}_{z^*z^*} \approx \frac{w_{k+1} - 2w_k + w_{k-1}}{\Delta^2}, \tag{A.15}$$

where a grid notation has been introduced: $w_k = \tilde{w}(z_k)$. The finite difference version of Equation A.3 is then

$$w_{k+1} + (\Delta^2 Q_k^2 - 2)w_k + w_{k-1} = \Delta^2 F_k \tag{A.16}$$

for $k = 2, \ldots, K - 1$. For $k = 1$ we have

$$w_2 + (\Delta^2 Q_1^2 - 2)w_1 = \Delta^2 F_1 - \tilde{w}_{bot}. \tag{A.17}$$

For $k = K$, our relation depends on the upper boundary condition.

(i) Lid:

$$w_{K-1} + (\Delta^2 Q_k^2 - 2)w_K = \Delta^2 F_K. \tag{A.18}$$

(ii) Radiation Condition:

$$2w_{K-1} + (\Delta^2 Q_k^2 - 2\Delta i Q_K - 2)w_K = 0, \tag{A.19}$$

where we have used Equation A.9 to obtain:

$$w_{K+1} = -2\Delta i Q_K w_K + w_{K-1} \tag{A.20}$$

with $\text{Im}(Q) < 0$. (Remember that our 'top' has been taken to be above all forcing levels; i.e., $F_K = 0$.) Equations A.16–A.19 correspond to an algebraic tri-diagonal system of equations and is solved using a standard Gaussian elimination technique which is described in the next section. The z^*-derivatives needed to calculate the wave fluxes are computed using centred differences:

$$w_{z^*k} = \frac{w_{k+1} - w_{k-1}}{2\Delta}. \tag{A.21}$$

A.4 Gaussian elimination algorithm

Define

$$a_k = \Delta^2 Q_k^2 - 2 \tag{A.22}$$

$$b_k = \Delta^2 F_k \tag{A.23}$$

for $k = 2, \ldots, K - 1$, and its natural modification for $k = 1$, K due to boundary conditions (Equations A.17–A.19). The matrix problem to be solved is of the form

$$\mathcal{A}w = b, \tag{A.24}$$

where

$$w^T = (w_1, w_2, \ldots, w_K) \tag{A.25}$$

$$b^T = (b_1, b_2, \ldots, b_K) \tag{A.26}$$

and the tri-diagonal matrix $\mathcal{A}$ is given by

$$\mathcal{A} = \begin{bmatrix} a_1 & 1 & 0 & 0 & \cdots & 0 \\ 1 & a_2 & 1 & 0 & \cdots & 0 \\ 0 & 1 & a_3 & 1 & \cdots & 0 \\ 0 & 0 & \ddots & \ddots & \ddots & \vdots \\ 0 & 0 & 0 & \cdots & 1 & a_k \end{bmatrix}. \tag{A.27}$$

We give next a simple algorithm to solve this linear system of equations, using Gaussian elimination.
INPUT : $a_k, b_k, k = 1, \ldots, K$
OUTPUT: $w_k, k = 1, \ldots, K$

/Eliminate elements below diagonal
For $k = 2$ to K do
    $t := 1/a_{k-1}$
    $a_k := a_k - t$
    $b_k := b_k - tb_{k-1}$
end

/Back substitution
$w_K := b_K/a_K$
For $k = K - 1$ to 1 do
$\quad w_k := (b_k - w_{k+1})/a_k$
end
Notice that w_k and b_k may share storage.

The above algorithm solves for the complex values of w_k at each level. In practice, it is more convenient to look at the amplitudes and phases where

$$amplitude(w_k) = \left((\mathrm{Re}(w_k))^2 + (\mathrm{Im}(w_k))^2\right)^{1/2}$$

and

$$phase(w_k) = \arctan\left[\frac{Im(w_k)}{Re(w_k)}\right].$$

A.5 Suggested inputs and outputs

The following are parameters which should be left open for user input:

(i)   Number of vertical levels (must be greater than 3).

(ii) Height of the 'top' level.

(iii) Amplitude of the wave maker at the ground (i.e., w_{bot}). Since we are dealing with linear problems, the most useful choices for w_{bot} are 0 and 1.

(iv) The choice of upper boundary condition (lid or radiation).

(v)   The real part of c; that is, the phase speed.[2]

(vi) The damping rate. This is included as the imaginary part of c which equals a/k, where a =the linear damping rate and k =the horizontal wavenumber. It is usually convenient to input both a and k, and calculate $\mathrm{Im}(c)$ internally.

(vii) The details of the interior forcing given by Equation A.5.

---

[2]In some problems one will choose the vertical wavenumber instead.

(viii) The basic profiles of T and u. There are, of course, many ways of entering such profiles. The simplest way is to input the values of T and u at specific levels, and to interpolate between these levels. Straight line interpolations are adequate for many purposes, though cubic splines are sometimes better behaved. Analytic representations (like those used by Lindzen and Hong, 1974, and Fels, 1986) are also possible. In the exercises, the basic state temperature is frequently chosen to be constant – thus obviating the need for the above choices.

The nature of the output depends on the available output devices, and the elaborateness of the program. Tabular output of the amplitude and phase of the wave fields and of the wave fluxes is perfectly adequate for most purposes, though graphical output is occasionally advantageous. As always, output should include a summary of the relevant input choices.

References

The numbers in square brackets at the end of each reference listing refer to the chapters in the present volume where the reference is cited.

Amayenc, P. (1974). Tidal oscillations of the meridional neutral wind at midlatitudes. *Radio Sci.* **9**, 281–93 [9].

Andrews, D. G., J. R. Holton, and C. B. Leovy (1987). *Middle Atmosphere Dynamics*. New York: Academic [5].

Bartels, J. (1928). Gezeitenschwingungen der Atmosphäre. *Handbuch der Experimentalphysik* **25** (*Geophysik* 1), 163–210 [9].

Benney, D. J., and R. F. Bergeron (1969). A new class of nonlinear waves in parallel flows. *Studies Appl. Math.* **48**, 181–204 [10].

Beyers, N. J., B. T. Miers, and R. J. Reed (1966). Diurnal tidal motions near the stratopause during 48 hours at White Sands Missile Range. *J. Atmos. Sci.* **23**, 325–33 [9].

Bjerknes, V. (1937). Application of line integral theorems to the hydrodynamics of terrestrial and cosmic vortices. *Astrophys. Norv.* **2**, 263–339 [7].

Booker, J. R. and F. B. Bretherton (1967). The critical layer for internal gravity waves in a shear flow. *J. Fluid Mech.* **27**, 513–39 [10].

Budyko, M. I. (1969). The effect of solar radiation variations on the climate of the earth. *Tellus* **21**, 611–19 [2]

Butler, S. T., and K. A. Small (1963). The excitation of atmospheric oscillations. *Proc. Roy. Soc.* **A274**, 91–121 [9].

Chapman, S. (1918). An example of the determination of a minute periodic variation as illustrative of the law of errors. *Monthly Notices Roy. Astron. Soc.* **78**, 635–8 [9].

Chapman, S. (1924). The semi-diurnal oscillation of the atmosphere. *Quart. J. Roy. Met. Soc.* **50**, 165–95 [9].

Chapman, S. (1951). Atmospheric tides and oscillations. In *Compendium of Meteorology*, Boston: American Meteorological Society, 262–74 [9].

Chapman, S., and R. S. Lindzen (1970). *Atmospheric Tides*. Dordrecht: Reidel [9].

Chapman, S., and K. C. Westfold (1956). A comparison of the annual mean solar and lunar atmospheric tides in barometric pressure as regards their world-wide distribution of amplitude and phase. *J. Atmos. Terr. Phys.* **8**, 1–23 [9].

Charney, J. G. (1947). The dynamics of long waves in a baroclinic westerly current. *J. Meteorol.* **4**, 135–62 [13,14].

Charney, J. G. (1948). On the scale of atmospheric motions. *Geofys. Publ.* **17**(2), 371–85 [12].

Charney, J. G. (1973). Planetary fluid dynamics. In *Dynamic Meteorology*, ed. P. Morel. Dordrecht: Reidel [5].

Charney, J. G., and P. G. Drazin (1961). Propagation of planetary-scale disturbances from the lower into the upper atmosphere. *J. Geophys. Res.* **66**, 83–110 [11].

Charney, J. G., and M. E. Stern (1962). On the instability of internal baroclinic jets in a rotating atmosphere. *J. Atmos. Sci.* **19**, 169–82 [13,14]

Defant, A. (1961). *Physical Oceanography*, Vol. 1. Elmsford, NY: Pergamon [11].

Eady, E. T. (1949). Long waves and cyclone waves. *Tellus* **1**, 33–52 [14].

Elford, W. G. (1959). A study of winds between 80 and 100 km in medium latitudes. *Planetary Space Sci.* **1**, 94–101 [9].

Eliassen, A., and E. Palm (1961). On the transfer of energy in stationary mountain waves. *Geofysiske Publ.* **22**, 1–23 [8].

Farrell, B. (1987). On developing disturbances in shear. *J. Atmos. Sci.* **42**, 2191–9 [13].

Fels, S. B. (1986). Analytic representations of standard atmospheric temperature profiles. *J. Atmos. Sci.* **43**, 219–21 [Appendix].

Ferrel, W. (1856). An essay on the winds and the currents of the ocean. *Nashville J. Medicine and Surgery* **11**, 287–301 [7].

Gill, A. E. (1982). *Atmosphere-Ocean Dynamics.* New York: Academic [1].

Glass, M., and A. Spizzichino (1974). Waves in the lower thermosphere: recent experimental investigation. *J. Atmos. Terr. Phys.* **36**, 1825–39 [9].

Green, J. S. A. (1970). Transfer properties of the large scale eddies and the general circulation of the atmosphere. *Q.J. Roy. Met. Soc.* **96**, 157–85 [2].

Greenhow, J. S., and E. L. Neufeld (1961). Winds in the upper atmosphere. *Q. J. Roy. Met. Soc.* **87**, 472–89 [9].

Hadley, G. (1735). Concerning the cause of the general trade winds. *Phil. Trans.* **29**, 58–62 [7].

Harris, M. F., F. G. Finger, and S. Teweles (1962). Diurnal variations of wind, pressure and temperature in the troposphere and stratosphere over the Azores. *J. Atmos. Sci.* **19**, 136–49 [9].

Haurwitz, B. (1956). *The geographical distribution of the solar semi-diurnal pressure oscillation.* Meteorol. Pap. **2**(5), New York University [9].

Haurwitz, B. (1965). The diurnal surface pressure oscillation. *Archiv. Meteorol. Geophys. Biokl.* **A14**, 361–79 [9].

Held, I. M., and A. Y. Hou (1980). Nonlinear axially symmetric circulations in a nearly inviscid atmosphere. *J. Atmos. Sci.* **37**, 515–33 [7].

Held, I. M., and M. J. Suarez (1974). Simple albedo feedback models of the icecaps. *Tellus* **26**, 613–29 [2].

Hines, C. O. (1966). Diurnal tide in the upper atmosphere. *J. Geophys. Res.* **72**, 1453–59 [9].

Holton, J. R., (1979). *An Introduction to Dynamic Meteorology.* New York: Academic [1,3,4,6,8,12,13,14].

Holton, J. R., and R. S. Lindzen (1972). An updated theory for the quasibiennial cycle of the tropical stratosphere. *J. Atmos. Sci.* **29**, 1076–80 [10].

Houghton, J. T. (1977). *The Physics of Atmosphere.* Cambridge: Cambridge University Press [1,3,4,6,8,12,13].

Hsu, H.-H., and B. J. Hoskins (1989). Tidal fluctuations as seen in ECMWF data. *Q. J. Roy. Met. Soc.* **115**, 247–64 [9].

Jacchia, L. G., and Z. Kopal (1951). Atmospheric oscillations and the temperature of the upper atmosphere. *J. Meteorol.* **9**, 13–23 [9].

Kato, S. (1966). Diurnal atmospheric oscillation, 1, eigenvalues and Hough functions. *J. Geophys. Res.* **71**, 3201–9 [9].

Kelvin, Lord (Thomson, W.) (1882). On the thermodynamic acceleration of the earth's rotation. *Proc. Roy. Soc. Edinb.* **11**, 396–405 [9].

Kelvin, Lord (Thomson, W.) (1887). Stability of fluid motion: recti-linear motion of viscous fluid between two parallel planes. *Phil. Mag.* **24**, 188–96 [13].

Kertz, W. (1956). *Components of the Semidiurnal Pressure Oscillation.* Sci. Rept. 4, Dept. of Meteor. and Ocean., New York University [9].

Kuo, H.-L. (1949). Dynamic instabilty of two-dimensonal nondivergent flow in a barotropic atmosphere. *J. Meteor.* **6**, 105–22 [13].

Lamb, H. (1910). On atmospheric oscillations. *Proc. Roy. Soc.* **A84**, 551–72 [9].

Lamb, H. (1916). *Hydrodynamics*, 4th edition. Cambridge: Cambridge University Press [9].

Lau, N.-C., and A. H. Oort (1981). A comparative study of observed Northern Hemisphere circulation statistics based on GFDL and NMC analyses. Part I. The time-mean fields. *Mon. Wea. Rev.* **109**, 1380–403 [5].

Lindzen, R. S. (1966). On the theory of the diurnal tide. *Mon. Wea. Rev.* **94**, 295–301 [9].

Lindzen, R. S. (1967a). Planetary waves on beta planes. *Mon. Wea. Rev.* **95**, 441–51 [9].

Lindzen, R. S. (1967b). Thermally driven diurnal tide in the atmosphere. *Q.J. Roy. Met. Soc.* **93**, 18–42 [9].

Lindzen, R. S. (1968). The application of classical atmospheric tidal theory. *Proc. Roy. Soc.* **A 303**, 299–316 [9].

Lindzen, R. S. (1971). Equatorial planetary waves in shear: Part I. *J. Atmos. Sci.* **28**, 609–22 [10].

Lindzen, R. S. (1973). Wave-mean flow interaction in the upper atmosphere. *Bound. Lay. Met.* **4**, 327–43 [8].

Lindzen, R. S. (1974). Stability of a Helmholtz velocity profile in a continuously stratified infinite Boussinesq fluid - applications to a clear air turbulence. *J. Atmos. Sci.* **31**, 1507–14 [13].

Lindzen, R. S. (1978). Effect of daily variations of cumulonimbus activity on the atmospheric semidiurnal tide. *Mon. Wea. Rev.* **106**, 526–33 [9].

Lindzen, R. S. (1979). Atmospheric Tides. *An. Rev. Earth & Plan. Sci.* **7**, 199–225 [9].

Lindzen, R. S. (1981). Turbulence and stress due to gravity wave and tidal breakdown. *J. Geophys. Res.* **86**, 9707–14 [10,13].

Lindzen, R. S. (1988a). Instability of plane parallel shear flow (Towards a mechanistic picture of how it works). *PAGEOPH* **16**, 103–21 [13].

Lindzen, R. S. (1988b). Some remarks on cumulus parameterization. *PAGEOPH* **16**, 123–35 [7].

Lindzen, R. S., and S. Chapman (1969). Atmospheric tides. *Space. Sci. Revs.* **10**, 3-188 [9].

Lindzen, R. S., and B. F. Farrell (1977). Some realistic modifications of simple climate models. *J. Atmos. Sci.* **34**, 1487–501 [2].

Lindzen, R. S., and B. F. Farrell (1980). The role of polar regions in global climate, and the parameterization of global heat transport. *Mon. Wea. Rev.* **108**, 2064–79 [14].

Lindzen, R. S., and J. R. Holton (1968). A theory of quasi-biennial oscillation. *J. Atmos. Sci.* **26**, 1095–107 [10].

Lindzen, R. S., and S.-S. Hong, (1974). Effects of mean winds and horizonal temperature gradients on solar and lunar diurnal tides in the atmosphere. *J. Atmos. Sci.* **31**, 1421–46 [9].

Lindzen, R. S. and A. Y. Hou (1988). Hadley circulations for zonally averaged heating centered off the equator. *J. Atmos. Sci.* **45**, 2416–27 [7].

Lindzen, R. S. and A. Rosenthal (1976).   On the instability of Helmholtz velocity profiles in stably stratified fluids when a lower boundary is present. *J. Geophys. Res.* **81**, 1561–71 [13].

Lorenz, E. N. (1955). Available potential energy and the maintenance of the general circulation. *Tellus* **7**, 157–67 [14].

Lorenz, E. N. (1967). *The Nature and Theory of the General Circulation of the Atmosphere.* World Meteorological Organization Monograph No. 218, TP 115 [5,7].

Manabe, S. and R. F. Strickler (1965).   Simulated climatology of a general circulation model with a hydrologic cycle. *Mon. Wea. Rev.* **93**, 769–98 [14].

Manabe, S. and T. B. Terpstra (1974). The effects of mountains on the general circulation of the atmosphere as identified by numerical experiments. *J. Atmos. Sci.* **31**, 3–42 [14].

Margules, M. (1890).   Über die Schwingungen periodisch erwamter Luft. *Sitzber. Akad. Wiss. Wien, Abt.* IIa **99**, 204–27 [9].

Margules, M. (1892).   Luftbewegungen in einer rotierenden Sphäroidschale. *Sitzber. Akad. Wiss. Wien, Abt.* IIa **101**, 597–626 [9].

Margules, M. (1893).   Luftbewegungen in einer rotierenden Sphäroidschale. *Sitzber. Akad. Wiss. Wien, Abt.* IIa **102**, 11–56; 1369–421 [9].

Margules, M. (1903). Über die Energie der Stürme. *Jahrb. Zentralanst. Meteor., Wien* 1-26. In English translation in Abbe, C. (1910). *The Mechanics of the Earth's Atmosphere*, 3rd Coll., 533–95. Washington, DC: Smithsonian Inst. [14].

Martyn, D. F. and O. O. Pulley (1936). The temperatures and contituents of the upper atmosphere. *Proc. Roy. Soc.* **A154**, 455–86 [9].

Newell, R. E., J. W. Kidson, D. G. Vincent, and G. J. Boer (1972). *The General Circulation of the Tropical Atmosphere and Interactions with Extratropical Latitudes*, Vol. 1. Cambridge, MA: M.I.T. Press [5].

North, G. R. (1975). Theory of energy balance climate models. *J. Atmos. Sci.* **32**, 2033–43 [2].

Oort, A. H. (1978). Adequacy of the rawinsonde network for global circulation studes tested through numerical model output. *Mon. Wea. Rev.* **106**, 174–95 [5].

Oort, A. H. (1983). *Global Atmospheric Circulation Statistics, 1958-1973.* NOAA Professional Paper 14, U.S. Government Printing Office [7].

Oort, A. H., and E. M. Rasmusson (1970). On the annual variation of the monthly mean meridional circulation. *Mon. Wea. Rev.* **98**, 423–42 [7]

Orr, W. M. F. (1907). The stability or instability of the steady motions of a perfect liquid and of a viscous liquid. *Proc. Roy. Irish Acad.* **A27**, 9-138 [13].

Palmén, E. and C. W. Newton (1967). *Atmospheric Circulation Systems.* New York: Academic [5].

Pedlosky, J. (1979). *Geophysical Fluid Dynamics.* New York: Springer-Verlag [1,11,12].

Pekeris, C. L. (1937). Atmospheric oscillations. *Proc. Roy. Soc.* **A158**, 650–71 [9].

Phillips, N. A. (1954). Energy transformations and meridional circulations associated with simple baroclinic waves in a two-level, quasigeostrophic model. *Tellus* **6**, 273–86 [14].

Plumb, R. A. (1977). The interaction of two internal waves with the mean flow. *J. Atmos. Sci.* **34**, 1847–58 [10].

Plumb, R. A., and A. D. McEwan (1978). The instability of a forced standing wave in a viscous, stratified fluid: A laboratory analogue of the quasi-biennial oscillation. *J. Atmos. Sci.* **35**, 1827–39 [10].

Rayleigh, 3rd Baron (Strutt, J. W.) (1890). On the vibrations of an atmosphere. *Phil. Mag.* (5) **29**, 173–80 [9].

Reed, R. J. (1967). Semidiurnal tidal motions between 30 and 60 km. *J. Atmos. Sci.* **24**, 315–17 [9].

Reed, R. J., and D. G. Rogers (1962). The circulation of the tropical stratosphere in the years 1954–1960. *J. Atmos. Sci.* **19**, 127–35 [5].

Reed, R. J., J. L. Wolfe, and H. Nishimoto (1963). A spectral analysis of the energetics of the stratospheric sudden warming of early 1957. *J. Atmos. Sci.* **20**, 256–75 [5].

Reed, R. J., M. J. Oard, and Marya Sieminski (1969). A comparison of observed and theoretical diurnal tidal motions between 30 and 60 km. *Mon. Wea. Rev.* **97**, 456–9 [9].

Sarachik, E. S. (1985). A simple theory for the vertical structure of the tropical atmosphere. *PAGEOPH* **123**, 261–71 [7].

Sawada, R. (1956). The atmospheric lunar tides and the Temperature profile in the upper atmosphere. *Geophys. Mag.* **27**, 213–36 [9].

Schneider, E., and R.S. Lindzen (1976). A discussion of the parameterization of momentum exchange by cumulus convection. *J. Geophys. Res.* **81**, 3158–60.

Schneider, E. (1977). Axially symmetric steady state models of the basic state of instability and climate studies. Part II: Nonlinear calculations. *J. Atmos. Sci.* **34**, 280–96 [7].

Schneider, E., and R. S. Lindzen (1977), Axially symmetric steady state models of the basic state of instability and climate studies. Part I: Linearized calculations. *J. Atmos. Sci.* **34**, 253–79 [7].

Sellers, W. D. (1969). A global climatic model based on the energy balance of the earth-atmosphere system. *J. Appl. Meteorol.* **8**, 392–400 [2].

Serrin, J. (1959). Mathematical principles of classical fluid mechancs. In *Handb. d. Physik* **8**(1), Strömungsmechanik I, Springer-Verlag, 125-263 [6].

Siebert, M. (1961). Atmospheric tides. *Advances in Geophysics* **7**, 105–82 [9].

Stommel, H. (1948). The westward intensification of wind-driven ocean currents. *Trans. Amer. Geophys. Union* **99**, 202–6 [11].

Stone, P. H. (1973). The effect of large-scale eddies on climatic change. *J. Atmos. Sci.* **30**, 521–9 [2].

Taylor, G. I. (1932). The resonance theory of semidiurnal atmospheric oscillations. *Mem. Roy. Meteorol. Soc.* **4**, 41–52 [9].

Taylor, G. I. (1936). The oscillations of the atmosphere. *Proc. Roy. Soc.* **A156**, 318–26 [9].

Theon, J. S., W. Nordberg, L. B. Katchen, and J. J. Horvath (1967). Some observations on the thermal behavior of the mesosphere. *J. Atmos. Sci.* **24**, 428–38 [5,10]

Thomson, J. (1857). Grand currents of atmospheric circulation. *British Association Meeting*, Dublin [7].

U.S. Standard Atmosphere Supplements (1966). Washington, DC: U.S. Government Printing Office [2].

Wallace, J. M. (1973). General circulation of the tropical lower stratosphere. *Rev. Geophys. Space Phys.* **11**, 191–222 [5].

Wallace, J. M., and F. R. Hartranft (1969). Diurnal wind variations; surface to 30 km. *Mon. Wea. Rev.* **96**, 446–55 [9].

Wallace, J. M., and V. E. Kousky (1968). Observational evidence of Kelvin waves in the tropical stratosphere. *J. Atmos. Sci.* **25**, 900–7 [5].

Wallace, J. M., and R. F. Tadd (1974). Some further results concerning the vertical structure of atmospheric tidal motions within the lowest 30 kilometers. *Mon. Wea. Rev.* **102**, 795–803 [9].

Welander, P. (1959). On the vertically integrated mass transport in the oceans. In ed. Bolin, *The atmosphere and the sea in motion – Rossby memorial volume.* New York: Rockefeller Inst. Press/ Oxford University Press [11].

Wilkes, M. V. (1949). *Oscillations of the Earth's Atmosphere.* Cambridge: Cambridge University Press [9].